Project

ARK

Project
ARK

AWAKEN FROM EXTINCTION

John B. MacDonald
and
Lee E. Frelich

SEQUOIA
PRESS

Illustrations: Bly and Rowan Pope

ISBN: 978-0-9982414-0-1

Sequoia Press / Minneapolis

Dedicated to the Earth's biodiversity – the real wealth in the universe.

FOREWORD

Climate change has gone from theory to reality. It's already showing up in the weather floating above your head: extreme events are increasingly *super-sized*. The release of over a trillion tons of fossil fuel carbon and methane pollution—that's one trillion hot air balloons of man-made chemicals since the dawn of the Industrial Revolution—is now spiking floods, droughts and heat-waves. No, it's not your grandfather's weather anymore.

Sure, the climate has changed before, but this time around we can't point to the sun or volcanoes or wobbles in Earth's orbit. We are the volcano, and the exquisitely calibrated conditions favorable for not only our survival, but the fate of all species on the planet, are being stressed like never before. Global warming, pollution, deforestation, overfishing, acidification of the oceans and overde-velopment have already seen the world lose half its animal population over the last 40 years. A 2015 report in the journal *Science* examining 131 studies predicts that one in six of the Earth's species will be lost forever to extinction if world leaders don't take effective action on climate change.

If anything, climate models have been conservative: the volatility and disruption we're already witnessing on a planetary scale exceeds even the most alarming predictions thirty years ago. And here's the thing: in spite of a robust and rapidly-increasing body of data, evidence, and scientific validation, considerable uncertainty remains. Put simply: we don't know what we don't know. There will be tipping points nobody predicted in advance, discoveries that no computer model could anticipate. The reality: we're running a dark, Darwinian experiment, not only on the atmosphere, oceans and cryosphere, but on Earth's ability to sustain and nurture all life.

Project Ark is a cautionary tale and a worst-case scenario; a reminder that we're not nearly as smart or forward-thinking as we like to believe. I pray it won't come to this, but fiction can be a powerful motivator. Will our better angels prevail? Will our kids and grandchildren find the courage and conviction to clean up our mess and make a clean-energy economy non-negotiable? I'm still hopeful for the future, but we would all be well advised to pay attention. Take nothing for granted.

Man and every other species on Earth has benefited from the fruits of Creation. Every form of life on this precious planet is counting on us to do the right thing.

– Paul Douglas
Founder and Senior Meteorologist, AerisWeather
June 14, 2016

PREFACE

When we first began studying how humans might best apply their advanced understanding of scientific principles to solve environmental problems, we quickly came to realize the importance of having both scientific and spiritual spheres working *together* rather than competing against one another. The need for such cooperation became even more apparent as we delved deeper into the question of who we are in terms of our specific duty of care towards our planet and its inhabitants. Our telescopes and space probes have allowed us to explore vast regions of outer space, gaining new insights into the nature and origins of our universe, yet these same investigations have made it increasingly clear that the greatest mystery of all is right here on our own planet earth—life. Whether life exists elsewhere in the universe may never be known. But for the moment, at least, and until evidence is found to the contrary, the presumption must be that we are alone.

This living planet of ours is like a beacon of light shining through the vast emptiness of space, and it is indeed humbling for us to ponder our stewardship role of this remarkable creation. But the beacon is beginning to flicker, and if we ever allow it to be extinguished, the loss would be irreparable. Earth has the unique ability to nourish all life found upon it and it is important for our survival that we respect all parts of this ecosystem. In recent times we have tended instead to value certain parts over others, the result being that the system has become unbalanced and the harmonious relationships that once existed among all living beings have become dangerously disrupted. A mass extinction, caused by human actions, is well underway. With the loss of biodiversity, the resilience of the Earth's ecosystems to major disturbances such as climate change, and subsequently the ability of the Earth to sustain human life, is being reduced.

In the not so distant future, we may find it necessary to live in domed cities or to colonize nearby planets to ensure our survival. Under such dire conditions, we may be tempted to conduct high altitude geo-engineering experiments to re-set the atmosphere, regardless of the unpredictability and extreme risks involved in such projects. Then again, we might simply decide to hunker down and make the best of a bad situation. But in any case the most tragic consequence of all would be the loss of life's unique relationship to our own planet Earth and in a broader sense to the universe.

We humans are resourceful by nature. When confronted by adversity, our imaginations and fortitude rise to the occasion. We thrive on challenges and often do best when the odds are greatest against us. Perhaps these same virtues will once again serve us well, but the road won't be easy. Life is resilient and if we reach out and extend a caring hand, it will respond in kind and flourish once again. But in the meantime, we must prepare for the worst by collecting and dry-preserving DNA from as many plants and animals as possible, including those still found in the wild and those held in captivity. The same advanced methods should be applied to genetic material already stored in frozen holding tanks. Then, if conditions improve at some point in the future, we will be ready to re-constitute the DNA, thereby bringing numerous species back from extinction into a more hospitable world. For even if we believe there is only a slight chance our planet will lose its ability to support life, we would be negligent to avoid taking action while we still have the resources and access to do so. The time to build the ARK of the 21st century is now while conditions are still favorable, rather than later, when many species will have vanished and resources to save the rest will be harder to come by.

As Warren Buffet put it recently:

"If an ARK may be essential for survival, begin building it today, no matter how cloudless the skies appear. If there is only a 1% chance the planet is heading toward a truly major disaster and delay means passing a point of no return, inaction is foolhardy."

1

They worked quietly, methodically, yet with the unhurried intentness of a team of line chefs facing a long series of incoming orders. Racks of test tubes lined the counters, along with large brown bottles of solution and several digital microscopes. A walk-in freezer stood against the back wall. The set-up had a makeshift look and much of the equipment bore the logos of high-tech companies that no longer existed. Still, there was an unmistakable energy in the air, as if something serious and vital was underfoot.

As he worked, Dr. James Grace would occasionally emit a muted "hmmm" without knowing it. His wife, Laura, seldom responded. She was busy with her own work. And she knew the brief purring sound was merely the audible expression that accompanied the appearance of a small technical problem. Dr. Grace was good at solving such problems. He'd been working with genetic material for decades at the San Diego Zoo, cataloging seeds and preserving the cell cultures of endangered species.

During those early years, Dr. Grace (often referred to as Noah in the popular press) had watched the number of endangered species grow, and eventually become overshadowed by the list of species presumed to be extinct. As the climate crisis intensified, he'd become outspoken about budget priorities that exploited the entertainment (and revenue) value of a few large and exotic animals while neglecting the investment required to maintain a wide-reaching repository of genetic material. When the authorities decided the time had come to downsize the zoo's preservation program even further, Dr. Grace had quietly established his own lab to continue his work.

From that time onward, his name appeared in the press less often, though as the environment continued to deteriorate due to rising air temperatures and massive plankton die-offs at sea, it became obvious to many that something more needed to be done to combat mass extinction, and rumors began to circulate that Dr. Grace was perhaps already doing it.

But having seen first-hand how short-sighted (and self-interested) the Federation could be, Dr. Grace had little interest in working with them again. He maintained his independence, though his funding consisted largely of under-the-table contributions from wealthy donors. He was seldom seen in public, and as a means of deflecting serious interest in his work and even his whereabouts, he began to cultivate the "crazy Noah" image that the tabloids had created.

When government entreaties to recruit him went unanswered, and efforts even to find him fell flat, Dr. Grace saw his reputation suffer yet another mutation, from that of a semi-mythic Orpheus who could tame the animals—or better yet, bring them back to life—to an "enemy of the people," who, according to government propaganda, knew many things that would be of great benefit to mankind, but kept them selfishly to himself.

As his public reputation declined and his life became ever more solitary, James drew strength from Laura, who kept him healthy, shared his vision, and helped him daily with his work. Steeped in the traditions of her Native ancestors, Laura never lost heart.

When Laura announced one day that she was pregnant, James was momentarily stunned, rather than overjoyed. It seemed to him that the clandestine life they were living, and the legal and perhaps even physical dangers they might meet up with at any time, were less than ideal conditions for childrearing. But that reaction was almost immediately replaced by one of glowing joy and anticipation. And after all, the work they were doing involved preserving the genetic material necessary to replicate, at some future date, plant and animal species that were disappearing from the landscape. How fitting, he mused, that they were also going to be nurturing a new member of their own species along the way.

When he shared these thoughts with his wife, Laura sighed, gave him an affectionate look, and replied, "Well, dear, you can think about it like that if you want to. The way I see it, we're going to have a baby."

During Laura's pregnancy, the work went on much as it had before. They had plenty of genetic material at their disposal; the problem lay in prioritizing species, and then re-encoding the ones they deemed most important in such a way that they wouldn't require expensive, high-tech refrigeration to preserve over long periods of time. Magnetic coding wasn't reliable in the long term; the Graces therefore went through the time-consuming process of sequencing and the resynthesizing a copy of the DNA of each species on an electrochemical micro-array using error-correcting Reed-Solomon codes. To increase the durability of the coding and protect it from degrading effects of oxidation and humidity, it was encapsulated in silica microspheres. The result was a synthetic fossil containing the genome of a given species, that, if kept in a cool, dark place, might well remain viable for eons!

Aside from its durability, such micro-arrays had a second virtue: they were extremely small. Millions of them could be stored in a shoebox. Dr. Grace had devised a scheme for identifying which was which, but keeping order among the files added further time and labor to the project. With some common species they found it expedient to store and index seeds in the conventional way using test tubes.

The couple sometimes mused on the likelihood that anyone would be around at a later date with the wherewithal to resuscitate the myriad of flora and fauna they'd painstakingly preserved. But they didn't allow such thoughts to undermine their daily efforts. Once a species was lost, it would be lost forever: the Graces were convinced there was no task more worthy of their talents and energy than preserving what they could of these remarkable life-forms.

From time to time Dr. Grace would make a trip to a remote location where the newly created fossils could be stored out of harm's way, making use of a business front in the guise of a detoxification firm. Each such journey carried a certain risk of discovery—small aircraft were not often seen crossing the desert wastes of central California. On the other hand, few observers were in a position to spot such a craft either, now that visibility was so poor and urban life had been confined to the interior of large, widely scattered domes. Better to make such transfers, the Graces reasoned, than to allow some government or rogue agency to seize all the fruits of their life's work in a single, well-timed raid. Their subterranean lab was equipped with rudimentary video monitors and warning devices, though the work they were doing was fraught with so many contingencies that a government raid seemed no more worrisome than a serious earthquake or extended power outage.

Keeping such fears in the background, the Graces took pleasure in their daily work of processing and preserving the genomes of individual species. The genetic material all looked the same, but each plant and animal they worked on conjured images and associations—taxonomic, medicinal, environmental. Working on a mule deer, Dr. Grace was reminded of the acorns the animals fed on in the fall. A pinyon pine would remind Laura of the tasty nuts she'd eaten as a small child. Handling the DNA of a short-horned lizard, Dr. Grace would recall the research he'd once conducted on the mysterious pineal gland in the reptile's forehead. Thus each day was an exercise in nostalgia as well a preparation for a future that neither of them could fully imagine. And each day, the baby Laura was carrying grew larger.

When the alarms went off, Dr. Grace was working on the leatherback turtle, *Dermochelys coriacea*. He was in a reverie of sorts, imagining the time when that species had roamed every ocean and commonly lived for a century or more. The flashing red lights caught his attention, and he looked up the monitor to see armed men advancing up the tunnel toward the lab. Lots of men.

"Laura," he cried. "They've found us!"

But Laura had seen the lights and was already loading her specimens into a large titanium box mounted on a wheeled carriage. Dr. Grace disappeared into the freezer and returned bearing another large container. Laura wheeled the box in his direction and he loaded the material inside.

Distant alarms had begun to sound, and a few seconds later they heard a loud crashing sound.

"That's the gate going down," Grace said. "It will hold them back for a while, though I'm sure they brought some plastic explosives. We don't have much time."

Laura ran from one table to the next, snatching the most important tubes and containers until the box had reached its capacity. Several test tubes fell to the floor with a crash, but she hardly broke stride.

Then they heard the explosion. Closer. "They're through the gate."

He glanced at the video monitor and said, "Too bad the tunnel didn't collapse. I think we have two minutes maximum. Let's get to the pods."

Closing the titanium box, he wheeled it to the far side of the lab, where he pulled back a rug to reveal a large trap door. Opening it, he negotiated the ladder and manhandled the heavy box into the first of two small transport pods sitting side by side in the external bay. After helping Laura down the ladder, he revved up the first of the pods, then settled himself into the second pod to get it going.

When he engaged the ignition, nothing happened. Not a sound, a crank, a flicker. Just an ominous silence. Laura stared at her husband.

"It's not working?" she said in disbelief, though she knew the answer.

"I'll get it going. But there's no time to waste. Yours is ready to go. Get in."

"Don't be silly, I'm not going anywhere without you."

"You must," he said, raising his voice slightly. "Think of the child."

"—I think you can fit," she said.

"—not possible..."

"I don't entirely remember how these things work," she said in an anxious voice. "I should have joined you more often on your flights. You'll have to show me."

There was a pounding noise from the far side of the lab, and a muffled voice could be heard.

"This is Captain Erik Reyes, FF. Dr. James Grace! You are in violation of the federal statute 3-39.40. Open immediately or things will get ugly around here."

Dr. Grace moved to the other pod, and as he passed his wife he took a small dreamcatcher necklace from his pocket and placed it in her shaking hands. "Just something I got for you."

"Show me what to do," she said.

The pounding on the door was growing louder. James lifted open the cockpit, climbed in, and said, "Watch this. Basically these two upper left—" he flipped a few switches, and the pod surged, ready to depart, "—and then to engage prior to acceleration, this one. Press this and the autopilot will take you straight to the airport."

Feigning confusion, Laura reached over her husband and pressed the button. She then slammed the hatch down, trapping her husband inside the active pod.

Rings of electromagnetic discs began to rotate as the pod levitated and began to pick up speed, with Dr. James Grace looking desperately, impotently back at his wife.

As he vanished, Laura could read her husband's lips: "Laura! Laura! Don't do this!"

"Goodbye, James," she replied softly. "Techihhila." (Lakota for 'I love you.')

Fastening the dreamcatcher hurriedly around her

neck, Laura slid the rug back over the trap door just as the door on the other side of the room smashed apart.

The ever-zealous Captain Reyes was in the lead. A shiver ran down Laura's spine as she caught sight of him. He hadn't seen her yet. But then he turned. Too late she slipped behind a pillar, as if by keeping silent she could keep her location secret a little longer. Then she began to run.

"There!" Reyes shouted, pointing in the direction of the sound. Laura could hear the oddly soothing sound of the stun-guns as the soldiers fired wildly in several directions. The gunshots were like a series of smooth stones plunging into the water at the best possible angle to avoid making a splash. Pfftt. Pfftt. Her heart racing, Laura wove her way through a collection of life-sized replicas of gorillas. A flake of plaster splintered off one of the apes as a dart grazed it. The statue teetered in the opposite direction and then toppled over, smashing to pieces on the concrete floor of the lab.

Though she knew running was futile, adrenaline and instinct kept Laura moving ahead. When she reached a wall of fiberglass rock just beyond the primates, she leapt through a gap that had opened in the long-disused material due to the explosion. She was outside! Brown smog covered the orange-red horizon. It was difficult to breathe in the scorching hot air. Laura leaned against a tall metal cylinder bearing the image of a blue whale and a sign:"Cryogenic Embryo Bank. Common Name: Blue Whale. Scientific Name: *Balaeoneptera musculus*. Status: Extinct 2055." Though her pursuers were closing in, Laura continued to run past dozens of similar tanks. Finally she reached a place where she could look out over the distant ocean. Pfftt. Pfftt. She felt a powerful stinging in her ribs under her left arm.

As she began to lose consciousness, Laura could see, off in the distance, a droidlike elliptical craft gliding silently out over the water.

2

It was a desolate scene. Treeless hills and plains stretched to the horizon in every direction. The landscape, molded by heaving frost, might once have carried an austere charm, but the acrid atmosphere, swirling dust, and paucity of ground cover had brought its already limited life-forms to the point of abject exhaustion. Islands of rock from the earliest periods of the earth's formation were surrounded by seas of dead and near-dead organic matter, their cheerless browns and grays foretelling the planet's imminent return to lifelessness.

The only thing Michael Costello heard as he approached the steel doorway in the hillside was the sound of his feet crunching across gritty soil and desiccated lichen. The only movement that caught his eye was that of wind-blown sand scuttering at ankle height and occasionally rising up into swirling orange dust-devils.

As he approached the door, Michael could read the handmade sign hanging above it: DOOMSDAY VAULT. It was obvious that vandals had long since penetrated the chamber. The door hung askew, tethered by a single hinge.

Michael was used to disappointment—in the last twenty years he'd experienced far more than his share of it—but his heart sank once again when he paused to reflect on how much damage had been done over time due to the sheer thoughtlessness of marauding thrill-seekers. It only compounded the harm brought about by the politicians seeking greater power and wealth at the expense of the planet and the life it struggled to sustain.

Michael swung the door open gingerly; it remained attached to the doorjamb, but seemed so decrepit he decided to give it a hardy tug. It came crashing down at his feet with a metallic clatter, sending up a thick cloud of dirt and dust. Michael looked around to see if the sudden noise had spooked any nearby rodents, but all was still. He wasn't surprised. Many of his colleagues were convinced that mammalian populations had become so rare as to be unsustainable. To most scientists, it seemed obvious that many rodent species were already extinct. Then again, few of them ventured out beyond the walls of the urban domes these days to check, and one thing the general decline in the earth's condition had made evident was how adept living creatures were at locating the micro-environments that could keep them alive, albeit in drastically reduced numbers.

Stepping around the door, Michael switched on his headlamp and proceeded into the tunnel, more alert to anything that might collapse on him than to the likelihood of encountering another living creature. The door to the inner chamber was intact, but the scene he encountered when he opened it was largely what he had expected: the laboratory was full of fractured instruments and equipment. The all-important seed storage units were rusted and bent out of shape—they wouldn't be of much use to anyone, and it was hardly worth investigating whether any of the genetic material they contained was still viable. Michael was quite sure that due to time, neglect, earthquakes, and lack of power supplies, the vault had stopped serving its intended purpose long before the beer-drinking revelers arrived. The image of desert "rats"—those hardy if basically insane humans who made lives for themselves in the harsh wilderness beyond the confines of the urban domes—partying on in the midst of technological collapse, was almost a cheery thought!

Michael had never been much of a boozer himself. He had never been wild in that way, perhaps because he carried a touch of wildness inside him that harmonized with the landscape and the living world. Or maybe it was his sense of "mission" that forced him to direct his energies relentlessly toward the enormous tasks facing civilization rather than dissipating them frivolously. If he'd been more outspoken about such ideals and ambitions, he might have been accused of delusions of grandeur. But Michael had kept his nose to the grindstone and quietly risen to the top, developing a reputation as a brilliant and entirely reliable scientific expert, if also a bit of a stuffed shirt—at least to those who didn't know him well. Though a few professional rivalries had developed in the course of his meteoric career, Michael's personal staff was loyal and admiring. They knew him to be slightly driven but also thoughtful, fun-loving in his own quiet way, and considerate. Those who didn't know him well—especially individuals of a cynical disposition—considered him a goody-goody and were suspicious of his close ties to the government. And it was hard not to be cynical, with rationing more stringent than ever and official pronouncements of "good times just ahead" so repetitive and unconvincing.

In the desolation all around him, Michael could see vestiges of what had once been—and what might yet be again, though the odds were getting steeper all the time. That's what kept him moving ahead on seemingly futile missions like this one.

Suddenly the shelves rattled behind him and several items crashed to the floor. Michael drew his gun instinctively as he crouched and turned, but he saw nothing. Then he heard the faint grinding of glass on concrete as an empty beer bottle continued to roll back and forth slowly, undecided where to come to rest.

Another minor earthquake had shaken the vault. Such events had become commonplace as the cryosphere of the planet contracted and the surface layers rebounded from the weight of vanished glaciers farther to the north. Unfazed by the event, Michael holstered the weapon and returned through the tunnel to the acrid world outside.

By the time the sunlight hit his face, he was beginning to feel faint. The plane was out of sight, over a hill a few hundred yards away. But in which direction? He drew a rebreather from his belt and took a few deep breaths to bolster his oxygen intake. There were paths leading off in several directions. The revelers must have come here fairly recently. Perhaps it was their favorite watering hole?

Michael chuckled. But just then another, less pleasant thought struck him: Perhaps the local rats were even now ripping valuable equipment off his plane. Michael's head was clearing and he now easily spotted the landmarks he'd noted on the way in toward the cave. He quickened his pace but negotiated the rise in the land cautiously, weapon drawn. The plane finally came into view— unoccupied, unmolested.

By the time he reached it, Michael was once again exhausted and slightly numb. He pulled the cockpit lid closed and reached once again for the rebreather, which was attached to his jacket with a piece of Velcrow. Then he dug into his pocket for the audio recorder. There wasn't much to report. "Dr. Costello, Michael, at Svalbard Seed Bank, Norway: No plant materials have been preserved. No equipment functioning. Not a living thing in sight. Evidence of recent human activity widespread."

He clicked off the recorder and sat looking out through the cockpit at the wasteland beyond. It had the desolate beauty of landscapes people used to admire in New Mexico and Utah—before such scenes became so commonplace, so widespread, so inescapable. Soon the entire world would look like this. Parts of Africa, Australia, and China already looked much worse. The technologies that had been developed to support large populations within controlled environments were the wonder of the age, but they weren't sustainable. Michael mused, not for the first time, whether some civilization of the future would look upon the ruins of these domed cities the way people today—those few who retained an interest in their own distant past—looked upon the extravagant Roman villas on the coast of England or southern France at a time when the surrounding countryside was the uncultivated playground of roaming bands of illiterate nomads.

Michael knew what the next steps would have to be if there was any hope of saving the planet. In fact, he was in charge of carrying out those steps. Project Blue Skies. That was the official name, but he and his colleagues had coined a more colloquial phrase for it: they were going to "blow up the sky."

He could feel hope draining from him as he mouthed the words to himself, almost in disbelief: "We're going to have to blow up the sky to save the earth."

As he revved up the plane and taxied to the end of the runway, Michael saw a red light flashing on the dash. He flipped the switch, a smile on his face.

"Miss me already, Alex?"

"With all my heart," a voice responded. "Did you find anything?"

"Well, the vault is still there, but it's a wreck."

"Sorry, man. I guess we knew it was a long shot."

"Yeah. But if we hadn't checked it out, we always would have wondered. Are we still on schedule? Any relevant updates on the timing of the solar flare?"

"The data remains consistent with our projections, Mike, tending slightly toward an accelerated launch. Right now we're just waiting for you to get back and sign off on the mission."

Michael typed "34.4780 N, 118.6281 W" into the Global Positioning Control Unit. The display read: ESTI-

MATED TIME: 5 HOURS 32 MINUTES.

Michael checked his position relative to the lab and said, "Tell them I'll be there by 2."

Alex clicked off the radio and spun around in his chair. The expedition had been a dud, but nothing disastrous had happened, which was a plus. Alex held his boss in high regard, in part because Michael had recognized the genius lurking behind his somewhat devil-may-care exterior. But Michael was such a go-getter, it seemed he wanted to do everything *himself*. After all, why should the director of a major program head out on his own into the wild a few days before a major launch? Before THE biggest launch that any of them would ever be a part of?

Many would have answered that Michael was just like that. He liked going "out there," and he probably thought he might notice something others would miss.

But there was another reason. Michael was leaving no stone unturned in his efforts to determine if there was any feasible alternative to the Blue Skies Project, because he wasn't entirely convinced the project would work. Finding a repository of healthy genetic material would be critical to the success of any Plan B. Alex knew this, and Michael knew that he knew. It was sort of a bond between them. Neither of them talked about it because it would have been unprofessional, unpatriotic, and self-defeating, and besides, there was really nothing much to say. But this shared concern—skepticism would have been too strong a word—colored their relationship and made even their harshest jokes acceptable to each other.

Alex knew Michael wasn't a showboater. Michael had been eager to promote the work of his younger colleague and others involved in the project. And Michael had noticed in Alex not only a great deal of talent, but also an obliviousness to matters of status and official protocol that brought Michael's own boyish side to the surface. Michael knew he needed that. He needed an environment in which to relax from time to time. On this count, too, they had quickly developed an intimate rapport.

3

Isabelle Nez, a striking young woman with high cheek-bones and straight, jet-black hair, gestured vaguely upward toward the gigantic metallic structure looming behind her. The building was impressive, covered with delicate exterior piping like a miniature oil refinery, but more elegant, sparkling clean, and devoid of soot and other residues. Having seen the tower, the children were getting a little restless, and she was making every effort to hold their attention by explaining what it was for.

"Thirty years ago," Isabelle said, "our government brought the brightest minds within the Environmental Control Agency together to think up dramatic ways to improve our deteriorating environment."

"What's *that?*" one of the students interrupted, pointing to a small wooden ornament dangling around her neck.

"That's a dreamcatcher," Isabelle said, "and I'll tell you what it's for when we're finished with our talk. But right now I want to ask you all a question." After resorting to the rebreather in her hand, she continued. "Who can describe for me some of the engineering projects that were put into effect back in those early days to return our environment to health?"

One girl, who appeared as if she'd just come from a beauty salon, volunteered: "They put up giant mirrors to reflect the sun's rays."

"Very good. What else?"

A rambunctious-looking boy piped up: "We vaporized space junk to create an energy shield."

"Well, it was something like that. That's a good answer. And what's different about the plan that the agency will be putting into effect soon?"

"Photosynthesis!" several of the children shouted out.

"Excellent."

Hoping the little girl had forgotten about the dreamcatcher, Isabelle continued her spiel, trying to sound hopeful. "Project Blue Skies will take climate-engineering to an entirely new level. It was recently initiated by inoculating the atmosphere with millions of tons of synthetic chlorophyll through the carbon nanotube space elevator you see behind me and several others like it around the

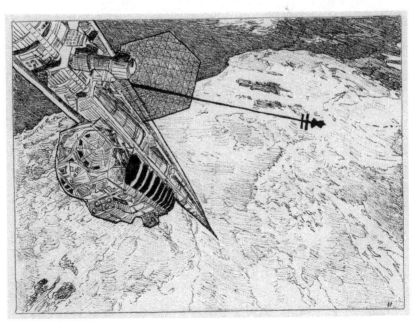

world. The next phase will take place in a few days, when the agency will launch a missile armed with an electromagnetic pulse bomb ...”

“To ignite the chlorophyll!” one of the kids shouted eagerly.

“Not exactly,” Isabelle replied with a smile. “There will be other launches in other parts of the world, all synchronized, all carrying e-bombs set to detonate at exactly the same moment. Can anyone tell me when that moment will be?”

Silence.

“Well, that was a trick question. I couldn’t tell you either. But the bombs have been set to detonate at precisely the moment when the massive solar flare you’ve all heard about on the news is predicted to reach its peak intensity. Our best scientists are convinced that the combined energy from the e-bombs and the coronal mass ejection caused by the solar flare will supercharge our atmosphere.”

One little girl raised her hand hesitantly.

“Yes?”

“My dad says it’s not going to work. He says that the solar flare is going to kill us all.”

“Well, a lot of very smart people think your dad is wrong—though I’m sure *he’s* smart, too,” Isabelle replied tactfully. She couldn’t resist adding, “—though I don’t think it’s very smart to scare your own daughter for no reason. What’s really going to happen is that we’re going to rearrange the chemical make-up of the atmosphere the same way these rebreathers improve the air we breathe. And this will be a good thing, not only for us, but for all the living creatures around us.”

“But there *aren’t* any creatures besides us,” one little boy said incredulously.

"Of course there are," Isabelle replied. "We just don't see them around because we don't venture out here that often, and they usually come out at night, when it's much cooler. But once we rearrange the atmosphere, temperatures will moderate and all sorts of good things will start happening. Nature is often like that. One good thing builds on another. And soon we'll recharge the atmosphere and begin the slow process of healing the earth."

"Or ending it!" another smart-aleck boy replied sarcastically.

Some of the children laughed; others looked uneasy.

"We can all see that life outside the domes can be pretty difficult," Isabelle said, gesturing toward the parched red hills surrounding them. "In a situation like this, when normal methods fail, more drastic measures are needed. But our talented scientists have come up with a very promising approach!"

"I think it'll work!" one girl shouted, caught up in Isabelle's enthusiasm.

"Yeah," several other students chimed in their agreement. Isabelle inhaled deeply through her rebreather and was about to continue when a man approached with purposeful strides across the parking lot.

"This event is terminated," he declared in a loud voice, waving his hands in front of him.

"What? I beg your pardon," Isabelle replied haughtily, startled but unfazed by the man's rude pronouncement.

"What do you think you're doing?" he replied gruffly. "These children shouldn't be out here. The atmosphere is deadly. And this is a classified area. You should know that." The man seemed inordinately upset.

"That may be," Isabelle replied, trying to keep her voice calm, "but I'd like to let you know that I checked the conditions before coming out here. And I also secured official permission before bringing my students on a field trip to this very important classified area." She flashed her guest ID tag, which was hanging around her neck alongside the dreamcatcher, at the unwelcome intruder.

"Well, consider your permission revoked," the man said, moderating his tone only slightly.

Isabelle was about to challenge the man's authority for terminating her visit when it dawned on her that she knew him. At any rate, she'd seen him before.

"I believe I'm speaking with the eminent Dr. Costello."

"That would be correct," the man said, eyeing her skeptically, perhaps eager to detect the slightest hint of sarcasm in her tone.

"I must say, you don't come across as quite so much of a boor in your television interviews."

The man finally cracked a wan smile and said, "Well, you shouldn't believe everything you see on the news. In any case, I'm merely concerned about the health of these children ... and the security of the state."

"In that order?" Isabelle replied.

"I can see you're a wiseacre," Dr. Costello said, frowning. "But the chemicals your students are breathing right now aren't the stuff of jokes. Their young bodies are more vulnerable than ours. If you don't leave immediately, I'm going to have to give you a citation and contact your superiors."

"Well, before you do," Isabelle replied sweetly, "I want you to know that many of these kids *do* believe what they see on TV. To them, you're a hero. What a great opportunity for these children to hear a few words about Project Blue Skies, which I've just been telling them about, from the man who designed it. Right, children?"

Cheers erupted. Dr. Costello looked confused. He flashed Isabelle a "you've got to be kidding me" look, but he realized that it wasn't such a bad idea, and he did his best to rise to the occasion.

"I've got a lot on my plate today, but I have time to say just a few words," he said. "But then you've really got to vacate the area. It's usually off-limits to civilians and it's not all that healthy out here. Your teacher is right, children. Atmospheric conditions aren't so bad right now, due to the—" Dr. Costello stopped himself. He didn't need to explain every detail to these kids. The teacher, with her penetrating eyes and calm demeanor, was beginning to fluster him. He was also strangely intrigued in spite of himself by a small scar he noticed on her cheek.

"So, to continue," he said, then added, "I'd like to take a closer look at that badge of yours, miss, before you go."

"Thank you so much, doctor," Isabelle said, a little obsequiously; perhaps she was laying it on a little thick. "A few words and then we'll be out of your hair—Scout's honor." She crossed her heart.

Michael flushed briefly as he watched Isabelle draw her finger across her chest. Turning to the students, he said, "Hi there, children. I'm Michael Costello."

"Hello, Dr. Costello!" several members of the class responded cheerily, almost in unison.

"I'd like to let you all know, first of all, that many people are involved in Project Blue Skies. I didn't design it personally, as your teacher here—" he paused, as if trying to remember her name.

"Isabelle," several of the children shouted with a giggle.

"Yes, as your teacher, Isabelle, suggested. Anyway, I'd like to explain the basic principles that will be at work in the project. The atmosphere has been over-rich in CO_2 for many years, and the effects have been devastating. You can see them everywhere. It may be hard for you to imagine a time when people walked freely from place to place in the open air, with green plants everywhere, and blue seas and lakes that you could actually swim in, but not so long ago that was indeed the case. Therefore, we've been saturating the atmosphere with synthetic chlorophyll for quite some time now. Our plan is to make use of the energy from an upcoming solar flare, along with an added boost from a few high-altitude detonations of our own, to initiate a process that will convert the excess CO_2 into harmless—"

"—oxygen and carbohydrates!" one of the boys called out.

"I see someone's been doing his homework. Who said that?"

One of the boys raised his hand.

Dr. Costello looked stunned. "Tommy?"

Confused, the boy said, "My name isn't Tommy."

Michael continued to stare at the boy for an instant, flustered and oddly devastated, before regaining his composure. When he recommenced his speech, his bravado had diminished.

"This tower you see behind you has been a part of the atmosphere-seeding process...and will also be active during the chemical operation," Costello continued in a monotone, waving an arm distractedly. "That's why you all have to leave, right now. The field trip is over."

"Thank you very much," Isabelle said. She had noticed his change in demeanor, though she had no idea what had caused it. "Now class, you heard what Dr. Costello said."

Just then one of the boys said, "I hear buzzing."

Dr. Costello was on the verge of departing but the remark caught his ear.

"You do?" he said, crouching next to the boy. "Where?"

"It's getting louder. I think it's coming from over there. It sounds like a drill. Now it's gone. It seemed far away, but somehow it also seemed loud. I didn't like that sound."

The boy seemed exhausted. He'd been outside the dome breathing bad air for too long.

"Can you still hear it?" Costello said. The boy shook his head. "Can anyone else hear it?"

No one spoke. All eyes were fixed on Costello.

"Listen, children," he said. "There are still lots of things going on out here that we don't quite understand. That's one reason we don't go out very often. And things that we don't understand sometimes seem scary. But most of the time, there's nothing to be afraid of. We're going to rest here for a bit and then make our way back to the base. It's right over there."

Isabelle was impressed that Costello had made an effort to listen to the boy. She was beginning to realize that there were two sides to this man, at least, and that the confident charmer on the screen wasn't entirely a facade.

The boy who'd heard the buzzing seemed less anxious now, though she wasn't. Costello was now standing stock still, evidently waiting for the sound to recur. The two stood motionless, straining their senses, trying to pick up anything unusual that the breeze might blow their way.

Though the wind had been gusting a few minutes earlier, it was now almost deadly calm.

"Look at how gray the sky is off to the northwest. And now I hear it again. It's getting louder. We've got to get moving," Michael exclaimed.

Isabelle was flattered that the eminent scientist had decided to accompany her class back to the safety of the base. The strange cloud to the northwest was growing larger, and there was now no mistaking the angry buzz that filled the air.

As the group moved forward, torn between the desire to hurry and the fear of leaving someone behind, a few small beetle-like insects began to land on the ground around their feet. Then, suddenly, a thick wave of insects descended with a roar.

"Keep moving. Don't panic!" Dr. Costello exclaimed, grabbing one of the boys who was lagging behind. The

arriving swarms cast a dark shadow over the landscape. The little party was in fact only a few steps from the entrance portal by that time, and the two adults ushered the children inside as briskly as they could without knocking them over, brushing away the little reddish bugs as well as they could. By the time the door to the outer portal finally closed, quite a few beetles were flying around inside. Trying to calm the children, and perhaps himself, Dr.

Costello said, "I've heard of such infestations, but I've never seen one before. These insects usually feed on dead organic matter floating out at sea. They only rarely venture inland."

Then he added, trying to sound cheerful, "I guess we're lucky we got to see some. But we'll all have to be fumigated before they let us back into the dome. This ought to be a lesson for us about the harsh and dangerous environment we've created for ourselves."

"We didn't create this world, you did," one of the children said, rather innocently.

4

Monitoring the remote, unmanned installations required for Project Blue Skies was largely a matter of keeping track of various readings—the chlorophyll output, the chemical makeup of the surrounding atmosphere, the humidity and temperature. Reports arrived hourly from China, Australia, the Siberian Republic, South Africa, and other far-flung places, and the data was ingeniously displayed on an enormous model of the earth that sat in the middle of the Environmental Control Agency headquarters. The data itself would be used to determine when the project would kick into effect, and with what degree of force, but the visual aid sitting where everyone in the control room could see it was both helpful and reassuring. As the chlorophyll levels rose, the simulation-earth grew greener, and the scientists watching it day by day naturally associated the changes with the greener earth that the project was designed to nurture—though not nearly so fast.

Teams of technicians visited the remote bases on a regular basis to perform routine maintenance, but occasionally Michael found sufficient reason to pay one of the nearer installations a personal visit. And if no rationale was readily available, he would sometimes venture out in any case, convinced that he might notice something important others had missed. Some members of the staff took this as a sign of mistrust and micromanaging, but those who knew him well recognized that Michael was a little bit restless, and also a little bit obsessed. He was carrying a lot of weight on his shoulders, and a half-day mission to Nevada or New Mexico often did him good.

One of the members of the core team at headquarters, Deborah Slater, had worked beside Michael for years. If she carried a torch for the unfortunate widower, she never let on: it wouldn't have been professional. But it was obvious to her that the exciting parts of Project Blue Skies were largely behind them, and Michael certainly knew it, too. The experiments required to get the chemistry right, the engineering involved in constructing the towers in far flung places, working with foreign construction firms with differing standards and methods—all of that was now behind them, along with the crucial analyses of meteorological effects that could make or break the entire operation. What lay ahead was the execution, which would be exciting indeed ... but only if it worked.

No wonder Michael felt the need to get away every now and then on a routine yet slightly dangerous mission.

One further element motivated Michael Costello to make occasional trips to some of the nearer unmanned sites, though he shared it with no one: he felt closer to his son, Tommy, when he was up in the air alone, crossing the desert wastes. It wasn't a feeling he could explain, even to himself, but it meant a great deal to him.

He had lost Tommy and the boy's mother in a freak environmental mishap not so dissimilar to the one Isabelle's class had just experienced. His wife had been a chaperone on a class outing. A small group had ventured only a few yards off the prescribed path to examine an exotic-looking plant—or so their classmates later conjectured. Tommy, like his father, had been drawn to the beauty of plant life even though there wasn't much of it around any more. As the group gathered around the plant, the ground had given way under their collective weight, and they'd all been buried in a landslide, never to be seen again. The landforms in the vicinity were so unstable that efforts to rescue them, or even to find the bodies, were considered too dangerous to undertake. Viewed in retrospect, the field trip itself had been deemed irresponsible, and court proceedings were initiated against the school, but Michael didn't join the plaintiffs. Life had become increasingly unpredictable. That was a fact. Yet it was important to take risks in an effort to become more familiar with the out-of-doors, even in its debased and perhaps irremediable state. Michael's grief had taken a different turn: he made a solemn vow to do everything he could to make the earth more fruitful, more varied, more hospitable—the kind of world Tommy would have loved. And he sometimes found himself wondering to *whom* that vow had been directed.

Yet on an emotional level, the tragedy was unshakable. It haunted Michael and was no doubt responsible for his harsh attitude toward the students he'd come upon recently. At times—and especially when airborne—he assuaged his grief by imagining that his wife and son were actually still alive, and had joined the desert rats who knew how to survive in the vast stretches of arid land that civilized dome-dwellers had no reason to think about. It was a crazy thought, and he knew it.

Michael was engaged in such thoughts while returning from one inspection trip, enjoying the cast of sunlight across the shapely dunes of the Mojave Desert. He could already see Mount Palomar Observatory, a small white pearl on a mountain ridge to the southwest, and soon Los Angeles would become visible, sitting like an enormous contact lens dropped by a careless god onto the rubble of a burning planet he no longer cared for. Nearer at hand below him, Michael could see occasional traces of abandoned highways and conurbations. He found himself humming the oldie tune "Love Hurts" and wondering idly if he was passing above the abandoned town of TwentyNine Palms, where country-rocker Gram Parsons had died a century earlier. In some places clusters of abandoned swimming pools had filled with dust and taken on the appearance of antique linoleum. The wind turbines in the pass north of Mount San Jacinto were cranking away as fast as ever: that was one source of energy the city could count on until the turbines themselves ceased to function.

Although the scene was desolate, Michael found it interesting, and almost inspiring, to consider what humans had once been able to create. Population levels had decreased drastically in recent decades, and he nurtured a hope that once Project Blue Skies resolved a few of the most pressing environmental issues, wise government and advanced technology might kickstart a sociopolitical renaissance. And if some sort of amnesty could be established, there was undoubtedly much to be learned from the rats who had succeeded in staying alive under the extreme conditions prevailing on and just below the planet's surface, hundreds of feet below him.

Michael was well aware that some of the ruins he was passing over were still inhabited, though it was difficult to tell which, and as he started his descent toward Los

Angeles, he knew he was entering the danger zone, where planes were low enough to be hit but still far enough away from the city to allow the rats who hit them to escape capture.

All sorts of stories circulated about the desert folk, many of whom had chosen to live under the harshest environmental conditions rather than accept a more comfortable life in the dome—a life that was accompanied by a variety of rations and strictures imposed by the Federation in the name of civil peace and wise resource management. It was rumored that some of the rats had actually been forcibly expelled from the domes, having been deemed socially troublesome and unlikely to be missed by anyone. Representatives of the Federation rankled at accusations of such oppressive policies and did everything they could to discredit them. It was true that the government controlled most media outlets; then again, who else was going to support them? Advertising was largely pointless in a society where there were few things to buy or sell except on the black market. Anyone who considered the situation objectively would see that the Federation was doing its best to keep the urban populations both informed and entertained.

Yet if life inside the domes was marginally comfortable, it was also numbingly dull, and ennui was widespread. The government pharmaceutical dispensaries were doing their best to keep people happy, though here, too, it was difficult to distinguish between nurturing good mental health and opiating the masses.

Was it any wonder that some citizens of Los Angeles idolized the desert rats, who had ingeniously reworked abandoned machinery and put it to new uses, jerry-rigged all sorts of energy-saving devices, and developed a variety of means to glean sustenance from the seemingly lifeless landscape? The young were especially enamored of these subterranean bands, admiring their love of freedom and their outlandishly ragged clothes while ignoring their lawlessness. One or two intrepid academics had even made attempts at ethnographic research, though their results were spotty and riddled with sympathetic bias.

Michael hadn't seen many desert rats face to face, and he was glad of the fact. They were said to be grotesque, with leathery skin, matted hair, and all manner of protective face- and headgear. His experience with them came mostly at the receiving end of long-distance potshots taken by small units of men who would appear suddenly from the shadows with shoulder-mounted missile launchers—who could say where they got them? Such attacks were seldom successful, but experienced pilots knew when to expect them.

The dome of Los Angeles was looming when Michael saw the first flash. It was a late attack, but those were the most dangerous. The planes were lower and the men responsible for such attacks were highly capable, having proven themselves adept at eluding the nearby dome authorities. He banked the plane at just the right moment so that the missile's guidance systems would lack sufficient time to react. As the projectile whizzed past, well off target, he caught a glimpse of men scurrying into the shadows of the foothills far below.

Now the dome was larger, it was immense—an impressive orb in the desert, exhibiting a golden luster that stood in glaring contrast to most of the lives contained within it. Within Los Angeles—and the other domes scattered around the globe were no different—an elite made the political decisions, allocated the resources, and sustained the delicate technological arrangements required to sustain life. Though everyone within the dome benefited, it was obvious that a disproportionate amount of the social and material fruits accrued to those who held the reins.

As Michael approached the portal, he once again saw the holographic banner, distorted and blinking above the dome: WELCOME TO LOS ANGELES. OUTSIDE TEMP: 130° DOME TEMP: 96° CAPACITY: FULL. He also took note of the stark contrast between the "swimming pool zone," as he called it, and the much larger "tenement zone." He was not one of those who admired or envied the desert rats—he knew how toxic the environment they were living in had become—but there were times when he wondered if those rats might hold the advantage in some ways over the masses of people he was cruising above.

As the plane approached the Environmental Control Agency port, Michael activated the intercom in the cockpit and said, "ECA-715, requesting approval to land. This is Dr. Costello."

"ECA-715, you're clear to land," a neutral-sounding, almost robotic voice replied. "Welcome home."

The doors of the dome's north flight deck were open when Michael completed his descent, and he brought the plane to rest on the runway without difficulty. A routine flight, not much accomplished; he was glad to be home.

But as he exited the cockpit, Michael was surprised to see what appeared to be two Secret Service agents standing nearby on the tarmac, alongside a trim, well-dressed Hispanic gentleman of medium height. It was President Reyes.

"Ah, Costello, I heard you were out at the elevator today. How are things progressing out there?" the president said, extending a hand as Costello approached.

Michael took the president's hand for a firm, brief

shake. "Everything is functioning normally. Chlorophyll levels are exactly where we want them, Mr. President. I haven't seen the data from other regions, but barring some unforeseen catastrophe, Project Blue Skies is good to go, whenever we get the signal from Our Mr. Sun."

"That's just what I wanted to hear. We're all counting on you to pull us through." The president flashed a confident smiled, as if he hadn't a care in the world, but Costello could see that the most powerful man on earth was worried. He was looking for reassurance, and Costello once again felt the heavy weight of the responsibility that the president had entrusted him with.

"I'll do my best, sir."

"Your best?" the president snapped back with unexpected vehemence. "We get one shot at this, Costello. Do or die. The pressure is on."

Then Reyes patted Costello on the back and laughed, as if he'd just stated the obvious—a gentle jibe between two seasoned pros.

Costello replied in kind. "I won't let you down." He paused, then added, as if to defuse any lingering anxiety the president might be feeling about the upcoming operation, "Incidentally, on my way in I spotted a couple of armed rats. Maybe seven miles to the northwest. They took a potshot at me."

"I'll send out a squad," Reyes said. "We certainly don't want any trouble from them tomorrow."

"My thoughts exactly, sir."

"Right. Carry on."

"Yes, sir." Costello nodded respectfully and strode off toward the terminal.

At least I didn't click my heels, he thought as he entered the building.

5

A few hours later the pressure was off: Michael Costello was sipping a drink at the Tipping Point Bar, luxuriating in the relative darkness and the random cacophony of young men and women unwinding all around him. He liked the place, and appreciated the double *entendre* in the name, though he was pretty sure that few of the bar's patrons stopped to think much about it. That was his job—to think—and besides, most of the planet's tipping points lay in the past, already tipped in the wrong direction. Though he wasn't much of a drinker, Michael visited the place often because it was nearby and offered an element of sociability and human contact even when he wasn't actually engaged in conversation. Too often nowadays, conversation soon swung around to the topic of "the project." Everyone wanted a bit of inside information. Everyone wanted to be reassured. Prior to the loss of his wife and son, such a place would have seemed empty, but now it satisfied a need. And perhaps the beer, too, contributed to the relaxing effect. He had spotted none of his colleagues that evening, and was content to drift and dream in solitude at the bar, nursing a lager. He turned when a portly man wearing thick glasses broke through the crowd and took the stool next to his.

"Alex," he said, a smile spreading across his face. "Fancy meeting you here. I'll buy you a beer!"

"I've got to tell you, that baseball cap looks ridiculous on you," Alex replied. "I hope you know that."

Michael shrugged. "Maybe I'm just trying to shed a few layers of my gleaming TV persona."

"Well, before you sink to the lowest levels of bad taste and personal oblivion—which would mean flipping the bill backwards—maybe you ought to take a look at this." As he spoke, Alex looked the other way and discretely slid a data chip across the bar.

"What is it? Now? I don't want to look at this."

"No, I think you do. For one thing, it's level 12, classified. Ever hear of Laura Strongbow?"

"Nope. Wait a minute! Was she the heroine in *Son of Paleface*?"

"Not funny. Is that your *second* beer?"

"You know I can't hold more than one."

"I also know you've always been sensitive to Native interests and issues."

"That's true. Which is why I know that most of our Native population has found it prudent to anglicize their names for one reason or another. To me, Strongbow sounds like a name out of a movie. But you also have a point. We're in a bar and I'm a little tired. And slightly drunk? Not really. In any case, I haven't reached the tipping point."

"Well, whatever. This Strongbow woman has been behind bars for a decade or two, so I guess she hasn't kept up with the latest indigenous trends. Anyway, I wouldn't have mentioned it if I didn't think it was important."

"What crime did she commit?"

"There are only rumors. But I think it might be worth talking to her."

Michael popped the chip into his personal device and scrolled a little, pausing to read a passage here and there.

"Wow!" he said finally. "Is this for real? Where did you get this? Reyes is all over it."

"Consider it a birthday present."

"Thanks, man."

The bartender had brought Alex, who was a regular, his favored synthetic brew, and Alex raised the pint glass, which had a decent head, high in the air.

"To Tommy."

"To Tommy," Michael echoed the toast as they drank. He was grateful to have Alex as a friend. No one else he knew would have felt comfortable raising the spectre of his son. Alex knew that Michael didn't want to forget. He wanted to move forward without forgetting the past.

At that moment the mood in the Tipping Point changed, as the huge screens lining the walls switched from a Dome League basketball game to a special report on the Blue Skies project. Few of the bar patrons had actually been watching the game, but now all eyes were fixed on the screens.

"And…you're on," Alex said jovially, if a little too

theatrically, and Michael pulled his hat down lower on his head.

The report began with an interview clip that many had seen before. The reporter, sounding a note of officious concern, said: "Many citizens are concerned about e-bombs exploding over their heads."

The camera switched to a close-up of Michael, exuding calm: "We're aware of these concerns. In fact, we share them. The safety of our citizens has always been, and continues to be, the number one concern at ECA." Turning to address someone beyond the field of view, he said, "Could you show the graphic, please?"

An image of a rotating earth appeared on the screen. The noise level in the bar increased slightly as some of the patrons, seeing nothing they hadn't seen before, returned to their private conversations. Others continued to watch as a simulated earth rotated beneath the arc of a rocket, which happened to explode just as a blazing yellow sun appeared above the horizon.

Michael heard someone a few stools down the bar say to his comely girlfriend, "These make-believe graphics are pleasant to look at, but they don't *prove* anything. They don't *mean* anything. The truth is, this may be our last night together. That's why I was thinking maybe you should come over to my place and..."

Outside, on the city streets, the same report was playing on immense billboards that were usually covered with garish advertisements for motor scooters, GMO fast food products, and synthetic opiates. A small crowd had gathered to watch on the sidewalk outside the Tipping Point. Isabelle Nez happened to be out, and she stood in a cluster with other pedestrians, though she'd heard it all before. Perhaps a few additional tidbits of information had been added to the presentation. But there, upon the big screen, was a "new" Michael Costello—one with whom she'd spoken earlier in the day. He looked different now, though he was describing, yet again, the details of the project.

"...the missiles will emit powerful bursts of electromagnetic energy into the upper troposphere at a safe distance of twelve kilometers. When the missiles explode, trillions of nanoparticles containing both chlorophyll-enriched chloroplasts and methane oxidizing bacteria will become charged and glow like floating specks throughout the sky. It ought to be quite a show..." Isabelle had gotten the feeling during their brief time together that Dr. Costello liked her. He appreciated her spunk, perhaps, rather than being put off by it. He had even directed his attention for a split second to the small scar on her cheek, as if he was pained to see that flaw on an otherwise appealing face. Well, what did it matter? She'd never see him again.

Inside the bar, Michael hunched down on his stool a little as he watched himself confidently delivering his rudimentary chemistry lesson for the fiftieth time. A formula now appeared on the screen.

PORPHYRINS + MAGNESIUM = SYNTHETIC CHLOROPHYLL

Michael heard himself say: "The result will be an immediate reduction in CO_2 levels, and this, in turn, will lower temperatures around the globe. It's the 'miracle' we've been working toward—but it isn't a miracle at all. It's the result of lots of investment, research, and hard work."

At that point someone switched the channel to a news station, where a panel of men and women was rehashing the possible risks and rewards of the undertaking in shrill tones, each panelist adept at breaking in at the earliest possible moment. Michael knew most of the speakers person-ally, and he had a hard time determining if the event was live or merely a repeat of one of the many discussions that had been held in recent months, most of them fashioned to excite public opinion, and hence get better ratings, rather than explore the various possible outcomes seriously.

"—electromagnetic pulse bombs! I don't care how safe he says it is, it's crazy!"

"—What other options are there? All the reports say the same thing, the domes in the United States and Austra-lia can be maintained for fifteen years at the most, and the ones in Russia, India, China won't last nearly that long."

"—I'd rather have fifteen goods years than e-bombs ex-ploding on my head. I doubt if I'll live that long, anyway!"

Someone sitting nearby threw an empty beer can at the screen. The throw was well off the mark, and the can clat-tered harmlessly to the floor in the shadows behind the bar. A waitress shouted to the bartender, "Bobby, turn that crap off! No one wants to hear those poindexters trying to outshout each other."

"Yeah," someone else shouted. "The Lakers' game is on Channel Five."

A few men in the back, evidently still interested in the political chatter, started chanting "Blue Skies! Blue Skies!" Soon the chanting filled the bar, not because people were that interested in the latest rehash of the news, but because it was fun to "get involved," and both the volume and the rhythm of the chanting sent their adrenalin racing.

The energy level in the bar rose as other patrons start-ed a counterchant: "Lakers! Lakers!"

"Blue Skies! Blue Skies!"

"Let's get out of here," Michael said.

"Good idea," Alex replied.

The two men slipped through the chanting crowd and returned to the street. "See you tomorrow," Alex said as he walked off into the night.

Michael headed in the opposite direction, hopping the sky-tram to his apartment near the old waterfront. Los Angeles had once been the nation's largest container port, but such commerce was a thing of the past. As the ice caps melted and the sea level rose, the docks had first been cleverly raised, and later extended inland and reconfigured. As trade declined, these sturdy, jerry-rigged structures were finally abandoned and smaller, more efficient docks and depots were constructed farther inland. By the time the city was enclosed, ocean-going trade had all but disappeared, though vessels still moved occasionally up and down the coast, hoping to evade the torrential storms that often moved along the Pacific Rim or appeared out of nowhere from the southwest. Over time, the ocean had become a matter of grave issue, rather than a poetic wonder—something to be avoided rather than sought out. It was dying, and during that process it was producing the sulfides that were rendering the atmosphere toxic to humans.

Yet the process was highly unpredictable, due to the stop-and-go nature of the glacier melt at the poles. Slow under-ice melting would lead to sudden fracturing of the sheets, sending enormous waves, but also masses of fresher water, south into the oceans, sometimes raising the water level several feet in a relatively short time. As the earth's crust rose in response to the shifting distribution of weight, earthquakes and volcanic eruptions would become more numerous.

Due to these and other factors, most of the container ships had been scrapped, but a few had been drydocked and ingeniously converted into condos. The "waterfront" neighborhood had developed a sort of shabby-genteel cachet, and there was no denying that it exhibited more character than the drab government-built apartment blocks in other parts of the city.

Michael was not the type to be living among and rubbing shoulders with the independent crowd—some would call it the "radical" crowd, disapprovingly—but he had a well-developed aesthetic sense. Though he was now in charge of a highly technical operation in which physics and chemistry loomed large, he had entered the world of scientific inquiry through the portal of zoology and botany, motivated by a sense of awe, which eventually spurred the desire to do anything he could to revivify the planet. From his apartment window, which faced the setting sun, Michael could see a few of the converted vessels immediately below him, and beyond them, the skeletons of abandoned loading docks stretching out toward the sea. For some reason, the scene reminded him of the underside of a turtle shell, where vestigial struts catalogued that creature's response to changes in sea level over the course of its development.

Having emerged from the sky-tram, he was now passing down the street alongside one of the more well-known commercial establishments, a former cartage barge that, because of its uncommonly old-fashioned nautical detail, had been retrofitted into a casino called the Jolly Dolphin.

Suddenly a man stumbled toward him out of the shadows beyond the streetlight.

"You're him...," the man said. He was obviously drunk.

Michael tried to steady the man.

"I'm afraid not. But I know who you're talking about. People say I *look* like him."

"*Oh*, it's you all right," the man insisted in a knowing way, as if to prove that no one could pull the wool over *his* eyes, drunk or sober. "You're the one that's going to end all this."

Michael was afraid the scene might get ugly, but having had his say, the man stumbled away, leaving Michael to ponder what he had actually meant by "end all this." Did he mean an end to the suffering and environmental degradation that had made everyone's lives so miserable, or an apocalyptic end to *everything*? In either case, it was hard to tell if the man had been angry or pleased by the prospect.

The brief encounter sent Michael's thoughts back to places he didn't want to return to. His job was to inspire confidence, among other things, and in the course of performing it he had almost convinced himself that the operation was inevitably bound to work. The science involved was relatively simple, it was true. But the scale of the operation would be vast, unprecedented. The nanotechnology had been devilish to develop and maintain, and the unknowns involved were considerable. The launch itself would be like running a deadly stretch of whitewater: even if Michael and his team knew the exact flow of the water and the precise location of every rock and ledge, that didn't guarantee they'd reach calm waters before going over. And despite a century of hard-core meteorological research, no one could really predict the flow of the jet stream around the globe for more than a few hours with any degree of accuracy. Imbalances and countercurrents might well upset the chemical mix in one region or another.

Tired of rehearsing a rerun of the arguments and counter-arguments inside his head, Michael's thoughts returned to the encounter earlier in the day. The little boy in the school group bore a striking resemblance to Tommy. That was strange, and also disturbing.

Back at his apartment, Michael dropped down on the couch. He toyed with the idea of enjoying a small tumbler of whisky but put the thought aside. Sure, his nerves were rattled, but after a good night's sleep he'd be himself again. He pulled a small toy buffalo from his pocket and stared at it.

Happy birthday, dear Tommy, happy birthday to you.

He set the buffalo down on the coffee table and looked across the room at a framed photo of a young couple, each with a hand on the shoulder of the small boy standing in front of them. He couldn't see it very well, but he didn't need to.

... gray days, all of them gone,
nothing but blue skies from now on ...

"Ahhhh!" he exclaimed suddenly, then grabbed at the buffalo on the table, knocking it over. He caught hold of the toy on his second attempt and slipped it back into his pocket. As he did so he felt something unexpected. The data chip Alex had given him!

Extracting it from his pocket, he looked at it this way and that, as if he'd never seen such a thing before. As if it were a toy or a miniature candy dispenser. He was tired; he was clowning. He knew what he'd have to do with the chip—look at it. Yet one more task before finally hitting the hay.

6

Over time, Isabelle had inured herself to the long line of storefronts in her neighborhood proclaiming their religious faiths in bright neon. It was a sign of the times. As opportunities for personal advancement withered, threats of annihilation grew, and political life took on the appearance of a shell game. People naturally began to direct their attention more seriously toward spiritual, and often brazenly apocalyptic, answers to questions of *meaning*. Isabelle was put off by the rhetoric and the hype, but she knew there were elements of truth involved, too. She also knew that these organizations received substantial government support, presumably because their local missions alleviated the government's welfare burden to some degree.

Isabelle herself had been raised in a quasi-religious environment by her adoptive parents. Though her stepbrother Abe had cast aside their parents' faith as an adult, Isabelle had continued to embrace it, nourished by Christ's emphasis on charity and compassion, and his nonjudgmental love of anyone willing to abandon the ego-centered path of self-aggrandizement. Isabelle had found her own path teaching young people about the wonders of God's creation—wonders that very few of them would ever see firsthand. And in her free time, she also put to use the expertise she'd gained during her botanical studies, sharing the therapeutic powers of plants with those who needed help. Such pursuits would probably have seemed earthbound and prosaic to her parents, and to many of the desperate people she passed on the street, but to Isabelle there was a quiet strength to be gotten from the earth itself, and she could think of no better cause to devote her life to.

On a typical afternoon, a bewildering variety of people would be coming and going on the street, and Isabelle could hear the voices of preachers delivering advice and admonitions in several languages as she passed by on the

way to the sky-tram. One kiosk displayed a video screen upon which a preacher was delivering a fire-and-brimstone oration to passing pedestrians, the gist of which might have seemed familiar to the Pilgrim Fathers.

And they heard the sound of the LORD God walking in the garden in the cool of the day, and the man and his wife hid themselves from the presence of the LORD God among the trees of the garden. But the LORD God called to the man and said to him, "Where are y-y-y-y-y-y-y-y-y-y-y-y-y...

The preacher's head began to jerk back and forth mechanically, and an elderly man sitting at a sidewalk table beside the screen began to fiddle with his laptop, frantically trying to get the video back on track.

As Isabelle passed she heard him say, "—son of a bitch." He looked up and she flashed him her stunning smile.

"Good morning, Maxwell," she said.

"Isabelle, my love, when are you gonna run away with me?"

"What? And leave all this?" She spread her arms with palms upraised.

"For you, darling, I would. Only for you."

Isabelle laughed as she continued walking.

A few minutes later she had arrived at the Jolly Dolphin casino barge, which looked far more dilapidated under the harsh light of day than it did at midnight, lit up with neon and buzzing with couples out on the town.

Isabelle knocked a few times at the main entrance, paused, then knocked again once. Finally she heard the deadbolt turn and the door was opened by a young man with curly auburn hair whose youth and guilelessness

were obvious beneath a well-groomed exterior roughness.

"Cameron," she nodded to the lad in greeting. "Is he here?" The man ushered her in and locked the door behind her before replying, "Down below."

She gave the man a second look before remarking, "Is that genuine facial hair?"

He blushed. "Go on, have your fun," he said.

"Actually, I think it's kind of sexy."

At that the young man began to stroke his chin thoughtfully, not sure how to respond. Then he said, "I'm all man, baby. You all just don't see it yet." The words sounded forced, if not entirely ridiculous, and as soon as he'd uttered them Cameron regretted having done so.

"Oh. I see…" Isabelle said over her shoulder as she moved down the steps and on through the main hall of the casino, passing men and women who were cleaning up and preparing for the coming night's gambling and festivities. She pushed through a door on the far wall and descended a set of steps to a lower deck, which had kept more of its functional nautical appearance.

As she entered the main hall, she sensed an atmosphere of heightened energy. Men and women were huddled in groups around tables, some of which were littered with maps and diagrams. A few looked up and acknowledged her presence with a nod or a smile.

Then Isabelle noticed the weapons. A few were stacked against rough-hewn wooden crates piled up against the wall. No doubt there were more inside the boxes, some of which had stencilling on the side: ALBACORE TUNA. Stepping over to a middle-aged man in front of a laptop, she said, "Hi, Lance. Hacked into any new government websites lately?"

"Well, I'm just trying to make myself useful," the man shrugged. Then he burst whimsically into song:

I hear you knockin'
But you can't come in

"Say, tell me," Isabelle said, gesturing toward the armaments, "where'd you get all of this?"

"It seems your brother raided a military compound last night. I figured you already knew about it."

She gave him a fierce, disapproving frown and continued past, hardly breaking stride. Lance rose from his seat and hurried to catch up to her.

"So you didn't know about the raid? You gotta talk to Abe, Isabelle. He's scaring the hell out of everybody. None of us signed up to go war, especially one we can't win."

Isabelle put her hand on the doorknob and then paused.

"Has my stuff arrived?"

"Yeah, it was dropped off last night." She continued to stare at him until Lance finally said, "Okay, I'll go get it for you."

"Thanks," she said, smiling slightly for the first time.

She opened the door and strode theatrically into an untidy office. The man sitting behind the desk did not look up. His dark hair and clean-cut appearance would have been considered handsome were it not for the intensity of his gaze and the latent anger that was bending his facial features slightly out of shape. "Morning, Sis," he said.

"What's going on?" she snapped back. When Abe didn't respond, she reached over the top of his computer monitor and hit the power button.

"Oh, come on, that was *March of the Penguins*. I love that movie."

An attractive blonde sitting at the adjacent desk said, "Isabelle, you know how Abe loves his pelicans."

"Penguins, Tina. They're called penguins," Abe said.

"I asked you a question," Isabelle said.

Suddenly all business, her brother replied, "I call it an insurance policy."

"Insurance for what?"

"For the end of the world."

"We've been over this."

"I'm not talking about the solar flare or the e-bombs," Abe said, grabbing a handful of reports and waving them in the air. "While those idiots out there figure out which government building to annoy next, I've been studying the bigger picture."

"Which is?"

"Fractions." Abe's girlfriend emitted a sudden giggle, catching them both off guard.

"Sorry," Tiny said. "In high school I was not good in fractions."

"You mean *grade* school," Isabelle couldn't resist the jibe.

"Be civil," Abe said, "and listen to me. Our water rations have been decreasing. Food supplies reduced. Earlier and earlier curfews on energy consumption. Where is all that lost material going?"

"They're trying to extend how long we can survive in this fish bowl. If you cut back on things, they last longer."

"Yeah. That's what they keep telling us. But that's not what they're doing. Little pieces go missing, a bit here, a bit there. Hardly anything to get worked up about. But over time these things add up."

"So what's your point?"

"Do you imagine, Isabelle, that the president and his yes men are going to allow themselves to suffer along with the rest of us if this grand project backfires? If the truth ever came to light, I'm sure they'd have an explanation about the importance of maintaining leadership—for the greater good, of course. The problem with their little plan is that someone *is* taking notice. Someone who doesn't buy their elitist rationale and has the means to take those valuable commodities back."

"Abe, what are you talking about? You're delusional. You're playing a very dangerous game for a meager jackpot. I'm beginning to wonder if you're out for a much bigger prize than a few kilowatts of stolen electricity."

"Like what?"

Just then the door opened and Lance walked in carrying a small white paper sack.

"Sorry to barge in like that," he said, sensing that the conversation he'd interrupted had been heated. "Here are the tinctures."

Isabelle took the sack and peered into it, shifting it to better expose the full range of its contents.

"I don't see the echinacea."

"It's all gone. The source has run dry," Abe said.

Isabelle looked at her brother in disbelief.

"It's just one more thing," he said. "Get used to it."

Then he continued in a slightly officious voice: "'The best way to take control of a population is to take away their freedoms a little at a time while keeping a smile on your face—one step forward, two steps back. Finally, you reach the point at which people stop noticing and the changes can no longer be reversed.' You know who said that, or something like it?"

"Let me guess. Adolf Hitler. But there are quite a few holes in that analogy, Abe. For example, you don't seem to realize that we're all living in a bubble, not because we want to, but because the world outside it is virtually uninhabitable."

"Yeah, yeah. So what does that make me? The Uni-bomber?"

"I would prefer the Pied Piper," Isabelle said. "But I guess that's for you to decide."

7

The next morning Michael woke up by degrees, slightly woozy and a little regretful. The martini had been *strong*. But he soon dressed and got the coffee going. He then reviewed the notes he'd taken while examining the material on the chip one more time to make sure he hadn't overlooked anything important the previous night.

Forty-five minutes later he was standing at the public entrance to the Los Angeles prison, which was located in a vaguely unhealthy quarter near the edge of the dome not far from the encroaching sea. Based on his rank and security clearance, he was admitted and provided with the information necessary to locate the inmate in question, and was soon traversing the building's slightly dank and dimly lit halls, following the directional signs. He arrived at the cell block he was looking for to find that there was a guard on duty, leaning against a wall in what seemed to be a condition of utter boredom, if not sleep. Michael cleared his throat.

"I would like access to the prisoner in B635," he said in a firm but courteous voice.

Startled from his reverie, the guard stood up straight and replied, as if by rote, "That's a restricted area."

"I realize that—" Michael said.

He was about to provide authorization when the guard, who had finally taken a good look at him, said, "Hey, you're the science dude from Blue Skies!"

Michael nodded, grinning faintly.

"What do you want with B635? She's a nutter, that one."

"I just have a few questions..."

Michael pulled up the sleeve of his shirt, exposing a bar code tattoo, which the guard scanned without further comment. There was a bleep and then a red light started blinking.

"Hmm. Access denied," the guard shrugged.

"But that doesn't make any sense."

The guard looked more closely at the scanner screen and offered an explanation. "I see. She's level 12."

"But prisoners can't be level 12. Documents, yes. People, no."

"Then maybe she *isn't* a prisoner. Or maybe she's a document. How should I know? I'm just telling you what the monitor says."

Michael glared at the guard for a split second, though it was obvious the man was making an attempt to be of service. Then, after a swift glance at the man's badge, he said, "Listen, Hank, I only want to see her for ten minutes."

"And I wanna fart rainbows, but it ain't gonna happen." The guard seemed to think this was funny. Perhaps he was enjoying his moment of authority, denying entry to a high-ranking scientist and media celebrity on unshakable grounds. All the same, the irreverence rankled Costello, but before he could think of a convincing response the guard pulled out his tabloid text-reader and suddenly became deeply interested in the day's news.

"I'll pay you," Michael said after a moment of awkward silence.

The guard looked up, eyed him for a few seconds and then said, "That's bribery. That's illegal...but I know you're legit, so let's say four liters."

"That's highway robbery! It's most of the day's ration!"

"So, you'll be thirsty today. I'm running a risk here, letting you in. I wouldn't even consider it but I've seen you on TV and know you're aboveboard."

Michael wasn't a haggler by nature, and he knew he needed to see Laura Strongbow. He sighed and held out his arm. The guard scanned his barcode and pushed a few buttons.

A robotic voice said: "Michael Costello, you have .24 liters of water remaining in your daily ration."

The guard then scanned his own barcode. The same voice informed him he now had 7.73 liters of water remaining.

"I was a hero a minute ago," Michael said.

"And now *I'm* the hero," the guard replied as he entered a few numbers on the security pad and slid the heavy door open. "Ain't it crazy how fast things can change? B section, turn left, end of the hall. You got ten minutes, hero!"

"Once I get there," Michael remarked over his shoulder as he headed down the long corridor.

He found the cell without difficulty. The door was open. A woman in late middle-age was sitting on the floor in the lotus position, evidently meditating. Michael hesitated to disturb her but knew he didn't have much time. Finally he said gently, "Laura Strongbow?"

After a brief pause the woman opened her eyes.

"Who are you?"

"My name is Dr. Michael Costello. I'm with the ECA."

The woman nodded. "Ah, yes. You're the one who's going to save us all."

Michael couldn't tell if there was sarcasm in the remark but he forced a smiled.

Laura rose to a standing position and said, "You don't believe it'll work."

"What gives you that idea?" Costello said. Laura stared at him as if she knew he was hiding something and was about to come clean. Finally he added, "I do have a few reservations. Nothing in life works out quite like you expect. We've got to plan for various contingencies. That's why I'm here."

"You want to know about the ARK."

"That's right. I've read your husband's work. It's amazing—"

"I'm sure he'd be delighted to hear it." Once again, Michael found it difficult to read the tone. Was Strongbow making fun of him?

"Does the ARK still exist? Do you know where it is?"

"What are we made of, Dr. Costello?"

The question caught Michael sideways. "I'm not entirely sure what you mean."

"Humans. What are we?"

"Cells, tissue, organs, intelligence."

"And what about the soul?"

"I'm not a religious man."

"So then where do you think the voices come from?" Laura asked.

"Voices?"

"The voice that reminds you in a whisper to do the right thing, even when it's hard. And the other one—the one that pushes you to do things you know you shouldn't. The one that says 'Go on, you deserve it, no one will know, it'll be all right.' "

Michael raised an eyebrow. *She is crazy*, he thought.

"Sorry, I've lost the thread. What are we talking about here?"

"I can tell you're not accustomed to discussing such things, doctor," Laura said. "We were examining the difference between the intellect and the soul. I was drawing your attention to the voices of conscience. And I was going to ask you if you'd ever thought about what happens to those voices when we die."

"They stop."

"They stop? Lights out?" She made a little gesture, as if she were pulling down on a string.

"Eventually, yes."

"Life. Just think of it. Most life-forms are extinct. We have a few insects, a few plants, perhaps an amphibian or two. Some bacteria and algae in dwindling amounts. Other than that, we're all that's left." She stared up at the ceiling light. Then she said, "We go on."

"Yes, I'm well aware of the predicament. Which is why I wanted to speak with you—"

"They're not really gone, you know."

"Please, stop talking in riddles. They've given me permission to speak with you for only a few minutes—"

"Because the world is coming to an end?"

"No. But we're running out of time, out of options."

"We?"

"Humanity."

"Always the hero."

"I'm not a hero. I'm a scientist looking for a better option. And I was under the impression that you and your husband were also scientists at one time. Can you tell me, is he still alive? Where is he? Is the genetic material safe?" He felt he was close to something important, but was growing exasperated by the vague metaphysical banter. Couldn't the woman just tell him what she knew? She was the one throwing obstacles in the path of survival.

"A better option? At this late hour?" She stepped closer to whisper in his ear. "Look to the sky, Michael. It's hidden in the stars."

"What's hidden? I haven't seen a star in fifteen years."

Laura looked genuinely stunned. "What? We can no longer see the stars?" Michael was about to snap back, "Where have *you* been?" but checked himself. He knew where she'd been. In a cell. He simply shook his head. No stars. She nodded, and a look of peace descended on her features. "Then I guess you'll have to blow up the sky."

Michael stared at the woman, utterly defeated. He'd already overstepped his limit. Just then his device rang. It was Alex.

"What's up?" Michael said.

"I'm at the ECA. You'd better get over here quick."

8

Michael arrived at the ECA headquarters to find a mob of protestors marching in front of the entrance carrying signs that read, WE DON'T WANT NO E-BOMBS! and STOP THE INSANITY! Others, in a slightly different key, advised SAVE YOUR SOUL — THE END IS NIGH. Michael passed through the security gates and parked in his assigned slot near the employee entrance. Alex must have been waiting in the lobby, for he appeared out of nowhere and moved with surprising speed across the parking lot to meet him.

"Where have you been?" he said as Michael got out of the car.

"I was at the prison. You know, checking up on the Strongbow material you gave me last night. It's a dead end."

Alex handed him a tablet and Michael began to scroll through the imagery as they walked.

"As you can see," Alex said, "the sun's magnetosphere is acting in ways we've never seen before."

"Well, one thing's for sure: the CME is imminent. It's a bit earlier than calculations suggested but not that far out of line. How much time do we have?"

Looking down at his watch, Alex said, "Our latest calculation is 2 hours, 42 minutes, 7 seconds, and counting."

"What? Is everyone here? Mobilize the team," he said.

Alex nodded as he raced off, once again carrying his chunky body with surprising agility. But Michael's order rang hollow in the air. Of course Alex had already mobilized the team.

Michael also hastened his step as he examined the data Alex had given him. The previous night's added research had worn him down, and his conversation at the prison had been discouraging, but he could feel his vigor returning as he reentered the labs where he'd spent so many hours in recent years. The faces were friendly, and the technology was familiar and still largely reliable, though signs of wear could be seen everywhere. And the task that lay ahead was obvious. Blue Skies had already been quietly in operation for several years, as the enormously tall space elevators dispensed a steady flow of nanoparticles into the upper atmosphere. Now that another critical element in the equation—the enormous burst of energy provided by the coronal mass ejection—was speeding toward earth, all that remained for Michael and his team was to determine precisely when to introduce the final bursts of energy provided by the e-bombs.

Passing through several labs, Michael arrived in a central, hangar-like space with an enormous 3D projection of the sun in the center. This model was used to convert data that streamed into the lab day and night from research centers around the world into something that could be seen and comprehended intuitively. To Michael, the orb had become an old friend, beautiful and full of life, but also fraught with unpredictability and menace. Its complex radiance and formal simplicity had pulled him out of a blue funk more than once since the death of his wife and son. At other times he was reminded, as he looked at it, of the furor caused by Galileo when he first reported that the sun had spots. At the time the official position, advanced by Aristotle and supported by the Church, was that the sun was a perfect orb, unchanging and essentially divine. The idea that it had blemishes of any kind, much less ones that *moved*, was heresy of a high order.

Michael had of late grown fond of the idea, unscientific though it might be, that the sun *was* divine. Tales of Apollo's chariot didn't interest him, but he found himself taken by the theories of Renaissance thinkers such as the Florentine Marsilio Ficino—theories that brought together astronomical observations, "ideal" geometric relationships, and cosmic notions of energy and goodness.

Such theories went nowhere, of course, considered from a practical or scientific point of view. In contrast, scientists were eventually able to draw correlations between those mole-like dots drifting across the sun's surface and various changes taking place on earth, in the stratosphere, and farther out in space. Researchers had eventually established relationships between sunspot cycles and the powerful solar flares that sent bursts of energy rocketing into space. It was messy science, like predicting earthquakes or avalanches, but it had improved markedly over time, and as the spectrographic instrumentation improved, scientists had finally become adept at predicting the appearance of solar flares with a good deal of precision.

That fact explained the role played by the digital clock at the base of the solar orb in the center of the room. It provided a countdown to the next anticipated coronal mass ejection, and also conveyed its likely intensity and direction. The clock was now at 2 hours, 22 minutes. The associated intensity rating was already higher than Michael had ever seen it. The direction was 45 degrees NE—a reading indicating that the CME would be heading almost directly toward earth. It was all pretty much as Michael and his team had predicted. It was just arriving a little earlier than anyone had anticipated.

An array of work desks, three deep, surrounded the orb. Most were occupied by men and women in casual dress—scientists and technicians. A few were vacant.

The success of Blue Skies was predicated on sustaining a few simple chemical reactions over an extremely broad area. The distribution of the particles involved in the project had been meticulously monitored from the beginning, and adjustments in flow had been made time and again to redress imbalances, and also to ensure that resources weren't being expended wastefully. In recent days technicians at the base had been active around the clock, monitoring weather patterns with special care. But during the final critical hours before the CME arrived, such remedial methods as adjusting flows, which were slow to take effect, would no longer be of much use. Though no one doubted that the template devised by the team would need only a few last-minute tweaks to be effective, everyone knew that the team might need to alter the pattern of where the detonations took place to compensate for slight variations in the concentration of chemicals in any given region.

Michael's ace in the hole was the input that could be provided, if necessary, from a station in Alaska run by the High Frequency Active Auroral Research Program. The station had been jointly founded by various academic and military organizations to study the ionosphere, well before the economic collapse that led to the formation of the Federation. It was now under the Federation's purview, like almost everything else, but as far as anyone knew, its mission remained focused on the broad swath of the earth's upper atmosphere that was susceptible to being ionized by solar radiation. The agency maintained a large subarctic base in Alaska equipped with a variety of instruments, not only to monitor but also modify the chemistry and electromagnetic properties of that region. Due to a number of coincidental factors, including the concentration of land masses in the northern hemisphere, many of the potential problems envisioned by the team could be addressed by means of the station's powerful tools and instruments.

Some would have called such adjustments guesswork, but they were the same calculations that sailors and river pilots had been making intuitively for centuries. Michael had been less troubled by vagaries in the execution of the project than by recent technical failures reported in several of the towers—one in Kazakhstan and another in Chile. Of equal concern was the spottiness of some of the weather data they'd been receiving. The programs that assembled the data into a meaningful shape at any given moment were designed to "fill in the blanks," as it were, in those places where information was absent or scanty. But that meant that an element of informed speculation was involved in all of their calculations—with the computers doing the guessing.

"Good morning, Dr. Costello," a young woman sitting at a desk near the aisle chirped.

"Good morning to you, Deborah," Michael replied. "Though I'm wondering if it will improve once we detonate a few warheads a mile or two above our heads." Such doomsday jests were the harmless engineering equivalent of the "break a leg" jibes common in the theatrical world, though there was an element of seriousness in Michael's tone that sounded a lot like gloom. He smiled wanly in Deborah's direction, and she tried to smile back, though she could see he was worried. Somehow, she had kept her faith over the years in the exhaustive analysis and preparation leading up to this moment, and in Michael's outsized contributions to them. She just hoped he himself would not lose confidence in them.

Michael continued on to the main control panel, which was surrounded on several sides by large screens. He checked various monitors, looked at one or two printed reports, and then started issuing replies to a few incoming messages. This was basically waiting time, and it was built into every project, every mission, like the extra time people once reserved for changing a flat tire on the way to the airport or making their way through security, back when civilians took to the air regularly. It was time in which to adjust to the first few things that went wrong. And it gave Michael the opportunity to examine the technical material Alex had given him more carefully.

The only abnormality thus far seemed to be the early appearance of the solar flare itself. The experts had detected weaknesses in the sun's magnetic fields but expected them to hold together longer than they did. When several zones snapped and then suddenly reconnected, they unleashed energy equivalent to thousands of hydrogen bombs earthward. A mere eight-and-a-half minutes after that initial observation, high-energy electromagnetic radiation reached the earth, temporarily disrupting communications around the world. Now, roughly two days later, an enormous mass of solar particles was due to arrive, traveling at five hundred miles a second. This would provide most of the energy the Federation needed to fuel the worldwide chemical reactions that would return balance to the atmosphere and revive the possibility that the earth could be healed.

Michael was beginning to flip on some of the screens that showed activity in other parts of the world when a little girl ran up to him and jumped into his arms.

"Bree, how are you?" he said, his face brightening.

"Uncle Mike! Daddy says it's time to fix the sky!"

"We're going to do our best, honey." He kissed her forehead and then put her down as Alex's wife, Stephanie, hurried over.

"Bree, I told you we're supposed to stay in our seats. We're very lucky that they even let us into the control room. Sorry, Mike."

Alex, who had been conversing with a colleague, now arrived and scooped up his daughter.

"Come on, sweetheart," he said. "I know which seat is the very best."

"Really?"

Stephanie touched Michael's arm. "Good luck," she said simply.

Michael nodded and then watched them climb to an upper tier where a few other nonessential viewers were seated in anticipation of the launch.

Michael returned his attention to the main console in front of him, which consisted of several large screens. He looked from screen to screen at the faces, most of which were looking directly at him, some from within the local lab, others from similarly active control centers in Novosibirsk, Melbourne, and other regional centers. He walked over to Deborah's desk to check some data. She was a brilliant and very dedicated single woman whom many felt should have risen to a loftier position in the scientific community. But she claimed she was content where she was and had never applied for a promotion or a transfer.

Michael conferred with several of his colleagues, who had devised the final routes and detonation sites of the e-bomb delivery units. No surprises there. The math was sound. Finally the moment had arrived to initiate the launch.

"Many thanks to all of you whose skill and diligence has brought us to this moment," he said. He double-checked a smaller monitor before continuing. "Russia, India, Australia, China, I appreciate how quickly you've all mobilized your teams. I have the president listening in online..." He hit a switch and the sun projection was replaced by a smaller but still impressive holograph of President Erik Reyes.

"Mr. President," he continued, "LaunchTeam USA and Launch Support Teams Worldwide are ready to proceed."

"Thank you, Dr. Costello. And thank you all. As you are all aware, the health of our world and the preservation of our way of life will depend on the success of the things you do today. We stand on the cusp of a new era. We have used the genius of human intellect to overcome seemingly impossible obstacles, thanks to you and your colleagues and predecessors extending back thirty years and more. We are now ready to commence the final and most important phase of Project Blue Skies, which will allow us once again to become masters of our own destiny. Make today count, as you stand on the shoulders of giants. God speed."

Michael scanned the faces staring at him expectantly from the screens, then addressed them one after another. "Team Russia, China, India, Australia? Are you ready to proceed?" The answers were crisply in the affirmative.

Turning to Alex, he said, "Do you see any anomalies, unexpected chemical imbalances. Any need to consider alternative detonation sites?"

Alex shook his head, then flipped his boss a casual thumbs up.

"Environmental Control System is GO for auto sequence start," Michael said.

Various team members responded in turn.

"Payload is GO."

"Propulsion is GO."

"Power is GO."

"Guidance and Control is GO."

"Communications is GO."

Finally Deborah, who was serving as a sort of overseer of the functional input departments, said, "Launch Integration. All five stations GO for auto sequence start."

There was something eerily unreal about the proceedings, Michael thought. It seemed like something out of a bad film, unbelievable and riddled with cliché. Yet everyone was doing exactly what they'd been trained to do. Hundreds of scientists and technicians around the globe had been working toward this moment for months, if not years. Perhaps that was what made it all seem so strange. The rehearsals and the genuine launch were identical. But the speculation was finally over. As for what would come next, nothing remotely like it had ever been seen beyond the confines of movie houses and research laboratories.

Michael shook off that train of thought and returned to business at hand. Raising his index finger, he said, "Prepare to start grid geoengineering ignition sequence. Initiate final phase countdown of Project Blue Skies."

A computer-generated voice responded: "Geo-Cyber Command authorizes your access. Welcome to Project Blue Skies launch command. Phase one initiation sequence. Proceed to Geomagnetic Solar Shield."

"Thank you. Request for confirmation we are go for launch ECA?"

The slightly robotic voice: "All systems GO."

"What is the current status of the coronal mass ejection?"

The computer voice responded with a sequence of technical specifications, followed by the words everyone was waiting to hear. "As predicted, the CME is approaching earth with an intensity consistent with successful implementation of the atmospheric reactions."

"People," Michael said, raising his hand.

At their various stations the team members simultaneously placed their hands on their identity scanners.

"The system is live," Deborah said.

"Confirm all inhabitants are in radiation blast areas and secure," Michael said. He should have done so earlier. At that point it hardly mattered. If conditions were right, the operation would continue regardless of the safety of individuals. Nevertheless it was good to know. Michael was relieved to receive the expected reply, and he continued with his litany of checks and counter-checks.

"Earth magnetosphere stable? Chlorophyll dispersion stable?"

"Checking resonance levels...sampling now complete. Nanoparticles have been successfully dispersed and have reached optimum saturation."

"Power up Alaska HAARP station as standby," Michael said.

"Check," a member of the staff responded.

"Alex, do you have a reading on the CME?" he said into his microphone.

"Sir, the CME will impact in just over seventeen minutes."

They looked at a computer screen that read: "Solar energy level 25×1025 Watts/Metre. CLASSIFICATION: CME LEVEL Z – WARNING, WARNING."

Alex said, "Whoooo ... it's coming."

Finally Michael said, "Charge space shield. Coupling missile EMP blast resonance with earth's magnetic field. Activate reactors. Initiate missile charge sequence on my mark. Five - four - three - two - one - mark!"

Down the line orders were given out regarding the silos and payloads. It was all routine, yet fraught with a delicate tension. The readings had to be correct, the seals and temperatures verified, the timing precise. Anxiety began to mount as the fateful moment approached.

"Count her down," Michael said.

"Ten, nine, eight, seven, six, five—

The room fell silent. Everyone stared at the screens in front of them. The rockets started to jiggle.

"Four, three, two, one…lift off."

"Let's save the world," Michael muttered to himself.

Laboratory personnel spent the next few minutes watching the position of the rockets as they sped toward their assigned payload drop destinations. Technicians had been assigned to each rocket, while a more general picture could be gained from the enormous globe in the center of the room, which had transformed itself once again, this time into a representation of earth.

Michael himself had a few minutes to relax and rehearse in his mind's eye how the next phase would unfold. Flipping a switch on his console, he said, "How's the weather, Deborah?"

"Cooperating," came the reply.

He flipped another switch, then inquired, "Hydrostatic, GC?"

"Trajectory's on target."

"Power source PWR?"

"Explosives primed."

Returning to the general intercom Michael said once again, "Count 'er down." The numbers seemed to echo across the cavernous room. One of the rockets slowly circling the representational orb disappeared at that moment over the Siberian horizon, but a moment later another one could be seen rising above Argentina.

"Payload in five - four - three - two - one …"

"Pull the pin."

Michael pulled the buffalo fetish from his pocket as he stared at the orb simulation. There were several intense bursts of light on the globe in the center of the room, each about the size of Labrador. Then a skein of lightning bolts shot across the globe's surface in every direction. Everyone's eyes were glued to the orb as the earth's atmosphere started to crackle and pop before assuming the appearance of a huge glowing ember.

These were digital effects designed by the simulation team, of course, but they were carefully cued to the data flooding in from every corner of the globe. If a missile's charge hadn't exploded, there would have been a shadow; if the energy network hadn't developed uniformly, the rendering would have been entirely different.

"It looks pretty much the way we'd hoped it would," Michael said, with a look more of relief than joy on his face. But he well knew that they weren't out of the woods, yet.

"I can't believe we're actually doing this," Alex said.

"Now the big question is, will the conversion of CO_2 proceed as we predicted?"

It was widely agreed within the scientific community that in order to sustain widespread life outside the domes, the atmospheric carbon content needed to drop to four hundred parts per million or lower. In recent years the number had never been less than one thousand.

The next few hours were excruciating. There was little anyone could do but wait and watch the monitors, both visual and statistical, while trillions of chemical reactions took place in the upper atmosphere all over the world. It reminded some old-timers of the exploratory missions during which a probe would disappear behind a planet and go silent for extended periods. There was no use examining the immediate data because irregularities and imbalances would be inevitable at first and would skew the results until the reactions became more widespread. The one meaningful thing that could be reviewed during those early minutes was whether the readings were coming in at all. Did the monitors still work, after the earth had been scorched by the fiercest electromagnetic wind in recorded history? The answer was yes.

Programmers had devised an algorithm to determine when the data had become widely meaningful. When that moment arrived, a series of loud beeps called Alex back to his desk. Reliable numbers were starting to come in.

"Call 'em out," Michael said.

"Okay. 990—"

That number was already an improvement over the CO_2 levels of the previous decade.

A few minutes later Michael said again, "Call them out."

"Okay. Not much is happening. Well, wait a minute. Here we have a 985. I see flickers of downward change. This is all good."

The minutes ticked by. A few of the monitors stopped reporting data, one in Quebec, another in South Africa. Still, the output could be considered strong, remarkable even.

"We're down to 970," came Alex's report. "And I think the rate of combination and reaction is accelerating."

As he watched the numbers, Alex had a sudden vision of something he learned in college. Nicolas of Cusa, a fifteenth century theologian, had argued that the very small and the very large end up being much the same thing. Alex couldn't quite recall what Nicolas had meant by the remark, but he felt he might now be seeing the principle in operation. Five million tons of nanoparticles, which are very small, had been dispersed over the entire globe, which was very large. One side of each nanoparticle contained a nano-produced synthetic chlorophyll which, when stimulated by energy from the sun (which was very large indeed) was converting CO_2 to O_2 and carbohydrate. On the other side of each nanoparticle, a nano-produced particle modeled after methane oxidizing bacteria was catalyzing the breakdown of methane molecules into carbohydrates that would then fall harmlessly to earth.

This very small but highly ingenious process was occurring trillions of times the world over. And that's why the CO_2 readings were going down.

But as the carbon content in the readings continued to drop, Alex was suddenly gripped with an irrational fear and began to sweat profusely. The numbers were supposed to drop. That was the point. But he began to wonder how far would they go. When the aggregate number reached 850, the lab personnel began to jump from their seats and cheer like football fans watching an improbable broken-field run. And even Alex, his fit of anxiety now behind him, gave the engineer standing next to him a big hug. Returning to his monitor he recited: "800. 770!"

"Come on 400!" someone shouted with ridiculous enthusiasm.

Anyone who had looked in Dr. Costello's direction as the results came in—and many did—would have seen him staring at a small plastic buffalo, deep in thought. Surfacing from his revery, Michael looked up once again at the data that had accumulated on the screen and then began to pace back and forth.

"Is Alaska HAARP reactor ready to go PROP?" he asked Deborah, who was standing at her desk nearby.

"It's ready if we *need* it, sir," she said, failing to conceal how unlikely she thought such an exigency would be. But after pacing for a few more seconds, Michael said,

"Prepare to initiate Alaska HAARP reactor for additional charge."

"Mike. Why?" Alex almost gasped.

"With all due respect, sir—" Deborah began. But Michael cut her off.

"This is our one and only opportunity. Such a CME won't occur again in our lifetimes, and if it did, we wouldn't have the resources to make use of it the way we're doing now. We can't take any chances that the conversion of CO_2 might fall flat. Right now I'm wondering how the energy beams from the luna ring panels are holding out."

"The energy is still coming in strong," Alex replied. "Those panels have proven remarkably durable. We're losing energy to the thickening atmosphere but nothing serious."

"So we can keep all the domes at full power while also energizing the HAARP reactors. Fire it up to 2.4 MHz and give me a 20% hold."

"But Mike," Alex protested, "there isn't any evidence that the rate of reaction is diminishing. The results so far are pretty consistent with the models we've been running. This is more or less the result we wanted."

"I'm not so sure," Michael snapped. "And once we drop below the 'kindling' point, we're doomed. There's no way we'd ever succeed in reviving the process. The time to act is now. Deborah, fire it up to 2.4 MHz."

"Yes, sir," Deborah responded, giving the words as much confidence as she could muster. She had never seen Michael so possessed, as if he were piloting a plane that had lost all its engines.

By this time the CO_2 level had dropped to 650, and Alex was a little disconcerted to notice that the rate continued to accelerate. They weren't that far from the target rate of 400 parts per million. How the energy from

Alaska would affect the levels no one could say for sure.

"640...625," Alex announced the carbon dioxide readings with obvious excitement in his voice. The level of chatter in the control room rose slightly as word spread that the HAARP instruments had been activated. The next few minutes were excruciating. It would take time for the impact of the added energy to appear in the aggregate readings, but anyone who happened to look at the model of earth in the center of the room could soon see small red sprites begin to shoot skyward from the Arctic region, looking less like flames than tiny needles of fire. That was something new, and not necessarily good. Michael saw them and turned away from the orb toward a more conventional monitor that offered a sequence of genuine digital views of what was going on outside from outposts in several locations.

What he saw when the sequence reached the Arctic region appalled him. Large flares raced horizontally across the sky. Downbursts of flame erupted from fuming red clouds alongside sheets of steaming rain. Michael couldn't help thinking that it looked diabolical. More than that, it looked apocalyptic.

"What's going on?" Deborah almost shouted at Alex, her brow furrowed, her hair in disarray. "None of our models predicted anything quite like this."

Alex's enthusiasm had vanished, to be replaced by a monotone of professional wariness, but as he watched these waves of heat and flames move across the rooftop of earth a lump formed in his stomach. Looking at his monitors, he happened to notice that the level of seismic activity was also increasing in northern locations. There was nothing he or anyone could do to stop the chemical reactions that were taking place, but it was obvious some major factor had been overlooked during the many months of lab trials. Now two question loomed: what were the reactions, and how far would they spread?

A pallor of silence had spread across the control room. Everyone was waiting, speechless, for whatever came next. At that point Alex noticed that the CO_2 levels were beginning to rise again! Rather than announcing the new readings, Alex turned to Michael, who looked ashen, and said, "They're going up: 514 ... 537 ... 552 ... 575 ..." His voice trailed off as both men continued to watch the screen. It was a slow, steady climb—651 ... 673 ... 745 ...

Finally Michael broke the silence. "The most likely explanation is that when we fired up the HAARP, we overcharged the atmosphere. The methane oxidizing side of the nanoparticles may have become so hyperactive from electron transfer imbalances that instead of converting methane into a carbohydrate, they're oxidizing it into CO_2. Also, it's possible that during the initial stages of the process far more methane was released from the sea and subterranean sources than we anticipated, due to the heat." He looked around vacantly, then continued. "To make matters worse, there is some chance that artificial photosynthesis has been halted due to chemical imbalances. If that were the case, the net effect would be that CO_2 levels would return to where they were before we started. Or go higher."

"For whatever reason," Alex said, "that seems to be where we're headed. And it looks pretty obvious that the methane is being ignited by the overcharged atmosphere."

All eyes were on Michael.

"That's it. Call it," he said.

"Mike, we don't *really* know what's happening. We need to give it more time," Deborah said.

"Deborah, I can think of no possible reason why the CO_2 levels would reverse direction yet again and begin to drop. Can you?"

Without waiting for an answer, he continued. "We have failed to meet our objective, and we've unleashed a demon reaction the properties of which we don't understand. The only responsible course for us now is to admit defeat, and do everything we can to avoid inflicting further damage."

"No. We can't give up that easily. If we give it more time—"

"We're back at eight hundred, Deb. Eight hundred!" Turning to another colleague, he gave the same order.

"Call it, David."

Another set of expert hands keyed in the requisite orders, and facilities throughout the world that had not been privy to these discussions or the preliminary readings terminated their operations, no doubt with many scratching their heads in confusion.

In the general devastation among the staff, all eyes had turned to Michael. Then they all turned discreetly away.

"I'll inform the president," he said as he hastily exited the room.

* * *

Later that day, Michael Costello found himself at an unusual after-hours meeting with President Reyes. During the brief span of time since the detonations, he had gathered together enough data to arrive at some preliminary conclusions as to what had gone wrong, and he'd also devised a means to explain the fiasco in terms that the president might understand. To a colleague in the field he might have explained things differently, but at the confidential meeting Michael informed the president, not for the first time, that the nanoparticles with which the atmosphere had been saturated were intended to act like chloroplasts, turning CO_2 into O_2 and carbohydrate by means of an artificial process of photosynthesis. At the same time, the methane released from the earth due to the heat would be oxidized via a chemical process similar to the one that occurs naturally when bacteria oxidize methane. Michael and his team had developed sophisticated formulas based on careful observation of these naturally occurring processes, and had arranged to have the relevant chemicals present in the quantities required to set those reactions in motion on an enormous scale. The intended result was to transform both carbon dioxide and methane into harmless carbohydrate particles. As the particulate matter became heavier, the nanoparticles to which it was clinging would drop to earth naturally due to gravity, cleansing the atmosphere of the dreaded carbon dioxide and methane that had unduly retained the sun's heat and largely stifled life on earth.

The president listened patiently to Michael's little speech, but he didn't appear to be listening too closely. When Michael had finished, the president said, "I don't need to know what was *supposed* to happen, I need to know what did happen, and *why* it happened, and, most important of all, what we're going to do next."

"In brief," Michael replied, "the materials we employed to produce the reactions were based on a specific level of energy, much of which was being provided by a solar flare of unusual intensity. If it had been possible to generate sufficient energy ourselves, we might have proceeded with Blue Skies years ago, when there was less CO_2 in the atmosphere and conditions weren't so bad. But it's extremely difficult to determine the force of a solar flare with precision before it actually arrives, or to mitigate it if it proves too intense."

"So you're telling me, now, that this entire grand project, utterly essential to the future of our planet, has been based on guesswork?"

"I'm telling you that we were aware of the potential danger, and were prepared to make adjustments to the energy inputs we *could* control to arrive at the desired result regardless of the strength of the mass ejection."

"And you made those adjustments, I presume."

"We did."

"And?"

"They failed. Otherwise put, I miscalculated. In fact, the end result of the project might well be to increase the CO_2 levels worldwide."

Michael looked at his shoes. President Reyes was silent. Michael wasn't sure whether the pause was to allow him further time to stew in his own ignominy, or because the remark was too devastating for the president to fathom or accept.

"You're saying the air quality is worse than when we started?"

"It seems likely. Reactions are still taking place in many parts of the world."

"Well, then. Where do we go from here? What's you plan?"

"There is no plan, sir."

The president raised his voice for the first time. "There has *got* to be a plan. Therefore, there is a plan. I'm not sure the man in the street has the foggiest idea of the chemistry involved in this dramatic display of fireworks, Costello, but he does know one thing: it cost a lot of money. Now, we can't exactly say that the project has been a total flop, can we?"

"No, sir."

"So we'll have to say that Blue Skies failed to achieve all of its objectives, and that the second phase of the plan will be implemented in the coming months once modifications have been made based on the data collected today."

"But there is no second phase."

"You'll think of something. But I would also recommend that you come up with some thoughtful and reassuring responses to the obvious questions you're going to be asked today. Say as little as possible, of course. Keep out of the public eye. Decline interviews. Meanwhile, I can have my speech writers work some things out for you."

"Thank you for the offer, Mr. President, but I don't think that will be necessary. I'm sure you are aware that while many citizens are satisfied with bread and circuses, increasing numbers have become well versed in the basic environmental sciences, what with all the trouble we've been having. I'm going to have to come up with a convincing story to tell *them*, too."

"Of course, of course. And one other thing, Dr. Costello. This new plan of yours—it had better work."

9

Throughout the day, Isabelle had allowed her students to watch updates on Project Blue Skies, though she considered it her first responsibility to shield them from ill-informed and alarmist reports, which wasn't easy considering how inflammatory some news outlets had become, and how beholden to government influence others obviously were. She had succeeded in exposing the children to brief media reports throughout the day, while keeping their attention largely focused on their daily assignments, as if the day were just like any other day, but with the added excitement of a world-transforming event ever-present in the background.

Isabella was naturally curious herself about the progress of the atmospheric reactions, and she made time to gather updates between lessons or when the kids were working together in groups. The snippets she gleaned were not encouraging.

It proved to be a long day, and she was glad when the last of the students said, "Bye, Ms. Nez, see you tomorrow," and ambled off home.

Isabelle picked up a stray ball and two hula hoops, returned them to the wooden toy box at the edge of the playground, and went inside to gather her things. A few minutes later she was heading home herself, too weary to notice the car that had been parked at the end of the block on the other side of the street for the last half hour. There was nothing extraordinary about the vehicle, except that two men were sitting inside, watching her every move through tinted glass.

* * *

As President Reyes and his entourage of Secret Service agents approached the prison checkpoint, Hank dutifully pressed the button that opened the gated entry. Without breaking stride, the loose pack of men continued into the prison. The president knew the way to the interrogation room: he'd been there many times. To the armed guards flanking the door he said, "Thank you, we'll take it from here." The guards stepped away with nothing more than a respectful nod, and two special agents took their place. Another agent opened the door for Reyes and followed him in.

Laura Strongbow was sitting with her hands folded in front of her on the table. She did not look up. She knew that Reyes, the old fool, would be staring at her with that intense gaze intended to frighten her into telling him things she had no intention of telling him. It was tiresome staring at President Reyes. But she didn't want to rile him up, either. She wanted to give him the impression that she was tired and might tell him what he wanted to know soon. That way, maybe he'd leave her alone again for a while.

Finally she lifted her gaze. Reyes had abandoned his icy glare, which served no purpose unless someone was looking into it.

"It's been quite a while, Laura," he said.

"Not long enough," she replied.

"I've got an idea. How about the two of us start over."

Reyes made a gesture with his finger and one his agents set a tablet onto the table where Laura could see it.

"That is a presidential pardon. I know I can't give you back the thirty years you've been in here, but I can give you a future."

Laura said nothing, though she moved her cheeks slightly so it appeared that she was thinking, or pouting.

"Aren't you tired of playing the martyr?" Reyes said, a little more angrily than he'd intended. "Your husband abandoned you. Does he really deserve the burden you've been bearing on his behalf all these years?"

Laura slid the tablet a few inches back across the table, using a single finger, as if she didn't wanted to be contaminated by it. Reyes's face twitched.

"We're running out of time, Laura," he said, his tone suddenly imploring rather than belligerent. "This dome and its citizens cannot survive without your help. Will you tell me where your husband hid the ARK? Won't you help me save the people of Los Angeles?"

"Spare me your speeches, Erik. Don't forget. I know you."

"Yes, I realize we've had our differences in the past, but that doesn't mean I'll always be the bad guy. I'm trying to save lives here—"

"—or gather more bargaining chips in your insatiable quest for power."

"Oh, that's a good one," he grinned. "You don't seem to understand what's going on. I already *have* power. And I'd be lying to you if I didn't confess that I enjoy having it. But soon the very notion of power, of administration, of making decisions and allocating resources, will be radically altered by the fact that there are no resources left to allocate. Not for me, or for you, or for anybody."

"And whose fault would that be?" Laura asked in mock curiosity.

"At this point it doesn't matter." Reyes pursed his lips for a moment before continuing.

"Listen, I want to show you something," he said. He swiped his fingers across the tablet, which was still resting on the table, then entered a few letters onto the keyboard. A holographic image appeared on the table of a woman strolling down a city street.

"Laura, that woman is your daughter."

The defiant visage vanished from Laura's face as if she had taken off a mask, and she stared, confused and enraptured, at the image on the table in front of her. Reyes let her flounder in her own emotions for a few seconds, and then he said, "Her name is Isabelle Nez. She teaches third grade. I've been keeping an eye on her for you, ever since the adoption."

"She looks like her father..." Laura's hostile tone had been replaced by an almost girlish wonder.

"I won't be able to save everyone, Laura, but I can save Isabelle. I can save your daughter, if you help me."

Laura continued to stare at the projection, mesmerized. President Reyes was about to repeat the remark when she smiled beatifically and said: "He who controls the key to life controls the world. It isn't in *your* power, President Reyes, to decide who lives and who dies. You're just another politician making promises you don't have the power to keep."

Reyes raised his hand on impulse, as if to slap her, but held back. He lowered his hand slowly, turned, and crossed to the door without a word, nodding on his way out to the agent who had accompanied him.

Laura continued to watch her daughter walk down the street. She saw her pause and say something to a street vendor. Perhaps it was a greeting; perhaps she walked along that street every day. Continuing on, Isabelle pulled her hair back, and Laura caught a glimpse of the dream-catcher hanging from her neck.

Just then the agent reached over and switched off the imagery. He picked up the tablet, and then he, too, exited the room without a word.

*   *   *

Following his interview with President Reyes, Michael Costello decided, on impulse, to see first-hand what sort of atmospheric havoc his Blue Skies project had wrought. Soon he would be required to produce a new plan, or something that looked like a plan. But so many resources had been devoted to Blue Skies that he simply could not imagine coming up with an inexpensive yet viable new scheme any time soon. If such a program or approach existed, they would already have implemented it. And there was none. They had given the planet their best shot, and it had not been enough.

So his plan, for the time being, was simply to examine the results of the previous plan. He knew the results were bad, so he decided to supplement his investigation with a bottle of whisky. That made the plan more appealing. Not the Blue Skies plan, but the plan of looking into the plan.

As he sat on the public sky-tram, Michael went over the day's events, his wandering mind returning repeatedly to the moment when he'd authorized the supplemental Alaskan energy-boost. Had that been a bold stroke of leadership or a failure of nerve? Was it a significant factor or a meaningless afterthought? The data analysts would soon come up with the answers to these and many other questions. If any of them were still alive.

Looking around him, Michael noticed the tram car was empty. He must be nearing the final stop. He had been swigging from his bottle discreetly, taking small sips, not only due to embarrassment, but also because he wanted to avoid passing out before he reached his destination. Now he took a heartier swig and rose to his feet unsteadily just as the tram was slowing to a stop.

"Ladies and gentlemen," a pre-recorded voice intoned, "this is the last stop on the line. Please exit the tram carefully, thank you."

Michael stepped from the car, spotted the nearby escalator, and made his way toward it, a little unsteadily. On the way down he took a swig, and then another, finishing off the bottle. He tossed it into a trash bin after emerging onto the street. The sound of the glass hitting the empty metal cans in the bin pleased him. He was on the verge of retrieving the bottle to toss it in again. Then it occurred to him that he had come to the edge of the city for quite another reason.

The neighborhood he was passing through was run-down. Many parts of the city were. The government housing had not been well maintained either by its landlords or its inhabitants, and the encroachment of the dome lent an atmosphere of congestion and even dankness to the area that residents closer to the more affluent city center didn't have to contend with. When he reached the dome's nearest exit portal Michael was stopped by one of the guards, who asked to see his authorization. He waved his badge in front of the man's impassive face and said, "Michael Costello, chief officer with the ECA. Experiment was a success…or a failure…I'm going outside."

Ignoring the cryptic remark, the guard scanned the barcode on Michael's badge and then stepped back to allow him to pass. Michael saluted and was about to proceed when the guard said, "You'll need a rebreather out there. The toxins in the atmosphere could kill you. I believe they're worse than ever after today's fireworks."

Michael grinned, a little idiotically. After all, what did this man know about the project? Nothing. No one had said the transformation would be instantaneous.

"Well, if I need a rebreather, perhaps you could *get* me one." Michael tried to put some official-sounding emphasis into the request, but he was having difficulty forming the words and wasn't sure he'd pulled it off. He was relieved when the man retreated into his little office-hut and returned a moment later with a rebreather, apparently a new one.

Now suitably equipped, Michael walked through the

several doors of the air-lock without difficulty and found himself facing a landscape of parched earth, dark reddish skies, swirling dirt, and a few weed-like shrubs. He could see the remnants of asphalt roads, and a few concrete foundations were also visible, though anything made of wood had long since been eaten by insects, and the wind-driven sand had buried most of the remaining signs of human presence. Roads—they were hardly more than paths—led off in several directions. Michael had been "out" many times before, but never from this portal, and he had only a vague idea where he was. He chose a secondary path that followed an incline. He was confident that it would soon drop down into a canyon. He wanted to lose sight of the city as soon as possible, and he didn't want to be followed or found any time soon.

That was all part of his marvelous plan.

Spectacular auroras swirled across the sky, which exuded an unnatural glow peppered from time to time with sparks. It would have been a beautiful display were it not so terrifying. Michael stared for a moment, mesmerized. He took a few steps up the hill in front of him and then, feeling weak, he pulled off his rebreather. Suddenly he wasn't sure if that was what he wanted to do. Get over the hill first. The toxic air was weakening him. He was already weak. He was drunk. He was acting stupidly. He drew the toy buffalo from his pocket and clutched it in his hand. Why was he out here? What was he trying to prove. He fumbled to reattach the rebreather while holding the buffalo but failed. Mustn't drop the buffalo. He went down on one knee, and the sand felt soft and warm. He began to sing—

Oh, give me a home where the buffalo roam
Where the deer and the antelope play.

And the skies are not cloudy all day.
And the buffalo play.
And seldom is heard a discouraging word ...

Then another tune rose in his mind.

Gray days, all of them gone.
Nothing but blue skies, from now on.

Then he turned over on his back and lay in the sand, staring up at the sky.

*   *   *

Following a familiar route, Isabelle Nez walked confidently down the streets of the Far End, a neighborhood so dismal that government investment was unlikely, while no amount of hype could improve its image. The rents were low, but so was the overhang of the dome. Services were minimal, and the local industries consisted largely of radiator-repair shops, gambling dens, and massage parlors. Yet Isabel knew that many elderly people on fixed incomes lived in the neighborhood, keeping out of sight for the most part. She knew some of them well, because they'd mustered the courage to visit her weekend clinic.

As she walked, Isabelle looked up through the dome to the sky above, which continued to spread flashes of lurid red through the deepening shadows of night. Turning up a side street close to the edge of the dome, she soon spotted an elderly couple out on their front stoop.

"Well, look at you!" she said with a smile. "Good to see you both."

She kissed the woman on the cheek and said, "Is that just the glare from the night sky, Mrs. Williams, or are you getting some of your color back? You look much better."

"She wanted to come out and see the sky," the man said. "I consider that a good sign."

"That was certainly an amazing display today, wasn't it?" Isabelle said. Mrs. Williams nodded and patted Isabelle's hand, which was still resting on her shoulder.

Isabelle squeezed her friend's hand affectionately, then stepped back and began fishing in her handbag. The light was poor but she finally located the bottle she was looking for. She handed it to Mr. Williams.

"This is lobelia. It will help her clear her lungs. The directions are written on the side."

"Risking yourself like that for us," Mr. Williams said. "You are a dear."

"Oh, I'm just a lowly schoolteacher. No one's watching me. Besides, we have to take care of our own, don't we?"

"They sure as heck won't take care of us. Bless you, darling," Mrs. Williams said.

"You take that like it says on the label. I think you'll notice some improvement soon, if not right away."

Mrs. Williams nodded as Isabelle waved a farewell and retraced her steps to the main road that led back to the center of the district. As she reached the corner she glanced toward the wall of the dome out of habit—no vehicles could be coming from that direction—and at just that moment she caught movement out of the corner of her eye. Perhaps a lizard had been scurrying across the dome's uneven surface, throwing a silhouette onto the street inside. Her curiosity was aroused: there were few creatures moving about these days amid the increasingly hostile environment beyond the dome. Walking over to the wall of the dome, which looked eerily like the plastic surface of a child's swimming pool, she wiped away as much grime as she could and peered out, only to see the whip of a lizard's tail as it scurried away.

She was pleased when the lizard turned to look back at her. But just then, in the distance beyond the creature, she saw what looked like the fallen trunk of a creosote tree. Or was it a human leg? It was hard to make out in the reddish gloom, but if it was a tree, the tree was wearing a shoe! Pulling her rebreather from her bag, she raced to the gate.

"I've just spotted an unusual lizard outside the dome, and I'm going to see if I can get a closer look," she told the guard, flashing her EDUCATIONAL RESEARCHER card. (Why hadn't she just told him the truth? Simple prudence, she guessed. She didn't know who was out there, or why, and there was no need to involve the authorities at this point.)

"At this time of night?" the guard replied, incredulous.

"It's a nocturnal species, seldom seen but never in the daytime. I suspect these unusual sky conditions have affected its vision. Different species see different frequencies of light—as I'm sure you know," she added in a vaguely flattering tone.

"Of course," the man said, beaming a little as he let her through.

Isabelle took a few quiet steps—the kind a biologist might take when trying not to disturb the local fauna—and then she raced up the hill toward the supine object, which rapidly took on the shape of a man as she approached.

It was Dr. Costello! She bent over him to check his vitals. Then she rummaged through her handbag, pulled out a small vial, and squirted a dropper-full of liquid into his mouth. Then she started in on the CPR.

"Breathe! Come on, BREATHE! Don't you do this, damn you! Breathe!"

She paused to take a few breaths through her re-breather, then continued the procedure until Costello's body suddenly erupted into spasms. The man gasped for air; Isabelle held down his flailing limbs.

"It's okay, you're going to be okay." She spotted the man's rebreather lying in the sand a few yards away and fetched it, then fastened it in place over his face.

"Can you hear me, Dr. Costello? Breathe slowly. You're going to be okay."

The man stared at her uncomprehendingly, then slowly nodded. Alive—barely.

As she collapsed in a sort of panicked relief, she suddenly felt two thick hands on her shoulders. She screamed as she was thrown sideways to the ground. The man—where had he come from?—seized her handbag, giving her an opportunity to jab him in the face with her elbow. A desert rat!

Surprised by her competent blow, the rat was knocked

back momentarily, but as she tried to scurry away he grabbed her leg, ripped the rebreather from her face, and stuffed it into his own bag.

Now he was sitting on top of her, pulling things out of the bag—dehydrated food pouches, a water bottle, a scarf. She struggled to free herself of his massive weight when she suddenly felt a blast of electricity race through her body. An instant later the man collapsed in a heap.

Isabelle, stunned once again by the swift turn of events, pushed the rat's suddenly inert body off her and pulled herself to a sitting position. Michael Costello was standing nearby on wobbly legs, weapon in hand.

"Is he ...?" She coughed, finding it difficult to breathe.

"He's just knocked out. I didn't want to risk doing too much damage to you. He probably has friends close by."

Isabelle retrieved her rebreather from the rat's bag, then moved to take the food and water, hesitated, then decided to leave them where they were. Michael had settled back down against a slab of concrete.

He said, "That's nearly a day's rations."

"I'll be given more tomorrow. He won't. Can you walk?"

Michael nodded but made no attempt to move. He was breathing slowly, and seemed to be enjoying every breath.

"He's just a rat," he said.

"So, he had a run of bad luck somewhere. It could happen to anyone. Once you're out, it's not easy to get back in."

Isabelle rose to her feet and offered Michael a hand. "Here, lean on me," she said. She got him up on his feet, put his arm around her neck, and started back toward the portal. By the time they reached it, it had begun to rain.

An hour later they were back at Isabelle's apartment, nearer the former waterfront. Michael was sprawled on the couch, still somewhat numb, as if suffering from heat stroke, but feeling slightly better. Isabelle had already given him several glasses of water, and now that it was past midnight she brought him another one, making a slight dent in her next day's ration. Before returning to the living room with the glass she added a few drops of a tincture to the water.

Michael rose to a sitting position as she approached.

"Here, drink this," she said. "We need to neutralize the toxins in your bloodstream and get them out of your system as quickly as possible."

Michael eyed the liquid suspiciously but was too exhausted to do anything but drink it. "What is it?" he said.

"You don't need to know. It will make you vomit, but I added a second drug to make it easier on your system."

Michael set the half-full glass down on the table with a clumsy thud.

"Drink it," she insisted. "We want to clear out those toxins." Then she added in a more playful tone, "You're not out of the woods yet. You could still die, you know, and I'd rather not have to figure out where to hide your body from the Feds."

"You'd hide my body?"

"Well, not me. I have friends who would do it for me."

"Probably the same friends who get you these potions."

"That's right. Now drink up. If you only drink half, you might just develop a hell of a stomachache for a few hours. I mean it."

"You don't look like the type of person who would be involved in such nefarious dealings. Do we know each other? I feel like I've met you before."

"Looks can be deceiving," she said, ignoring his question. "You don't look like the kind of guy who'd be wandering around drunk out in the dark beyond the dome on the day of a dreadful atmospheric test. What were you doing out there, anyway?"

"Just some testing."

"Oh, like testing whether an inebriated man can survive without oxygen in the desert? Doesn't seem very scientific to me. Where's the control group? Is that how Project Blue Skies was developed?"

The woman's questions were making Michael uncomfortable, but they were also lifting him out of his stupor. He finished the glass of water, as if to show her that he had understood its benefits and was not a total blockhead. Then he set the glass down on the end table—more delicately this time.

Isabelle disappeared into the kitchen and returned with a yellow plastic bucket. "I think you can figure out what to do with this when the time comes," she said. "It won't be long."

"Why were *you* out there?" he asked her suddenly.

"I saw something through the wall. I went to look for it."

Suddenly Michael half-stood and lowered his head while he vomited violently into the bucket. Isabelle handed him a cloth. After wiping his face Michael collapsed back on the couch.

"Did you find out what it was?" he said finally. Isabelle was impressed that her groaning guest had hung on to the thread of the conversation.

"Well, I got a pretty good look at it. I haven't had time to look it up, obviously." She took a few steps to the far side of the room, which was largely in shadow, and with-

PROJECT ARK

drew a book from the shelf.

"Are those books? It's obvious you have absolutely no respect for the law." There were hints of admiration in his voice.

Isabelle began flipping the pages, then stopped. She held out the book to show Michael a picture of a lizard.

"There. That's what I saw."

Michael looked at the photo, then he looked questioningly at Isabelle.

"You know that's not possible."

"I know. But I also know what I saw. Zoology is one of my specialties."

"Do they teach that in the Ancient History Department now?" Michael smirked.

She sat down in the chair across from him. They were becoming friends. They understood one another; she could tell by the tone of the banter they were sustaining. She rummaged through her handbag, which was leaning against the leg of her chair, and produced a small toy buffalo that she handed to him.

"I think you dropped this," she said.

Michael looked at it with a sort of fond detachment, and then said, "It was my son's. I promised him—I told him I'd find a way to heal the earth, maybe recover some of what we'd lost."

"And you have. Or you will, with the Blue Skies initiative."

"In fact, I won't," he replied, and a shadow of gloom fell across his face. "Blue Skies failed."

"What?" Isabelle exclaimed. She had heard the doubters and nay-sayers during the days and weeks prior to the implementation, but she had never allowed herself to imagine that such an expensive and broad-reaching—not to mention crucial—program could be a bust.

"What do you mean failed? I thought—"

"I mean 'failed' as in, 'we made the air worse, not better.' Look, I could be arrested even for telling you that. No one can know. It's highly classified—"

"I was so certain it was a sign," Isabelle said.

"What was a sign?"

"The lizard. I took him to be a messenger, a sign of better things ahead. A world full of thriving creatures."

"More likely it was a hallucination. No one has seen one of those for fifty years."

"Precisely my point. No one has—but I just did!"

Michael was about to reply when a ghastly look spread across his face and he retched again.

Suddenly there was heavy pounding on the door. Isabelle raced over, fumbled with the locks, then opened the door to a man holding a sawed-off shotgun in one hand.

"Good evening, Buck." Isabelle said, unperturbed. "What are you doing up so late?"

"Everything okay? Cameron said he saw some government guy—"

Isabelle opened the door wider so Buck could see Michael groaning quietly on the sofa.

"Our favorite scientist isn't feeling so well."

"I didn't know you two were dating." Buck said, looking at her quizzically.

Michael sat up straight when he saw the massive size of the man standing silhouetted in the door frame. Turning to address Michael, Isabelle said, "Buck here may be your biggest fan."

Buck moved with surprising quickness across the room and extended a hand, which Michael shook briefly. Buck glanced at the bucket. Michael stared at the shotgun.

"Whoa, Dr. Costello! I see you've been drinking. Well, listen up. If you hurt my girl here, I'll break your neck."

"Okay, noted. Thank you." Michael flashed a sheepish grin.

As he left the apartment, Buck said, "I'll leave my door open. Holler if you need me."

"Thanks, Buck." Returning to Michael's side, Isabelle said, "He's my guardian angel."

"That's one big angel. Can he fly?"

Just then Michael's device chimed.

"I'd better take this," he said. "Excuse me." He rose a little unsteadily and walked over to the bookshelves.

Isabelle watched him. His back was turned, his shoulders hunched. "What's going on?" she heard him say in a worried tone, louder than he'd intended, no doubt. Then he said, "Okay. I'm on my way."

"Duty calls?" Isabelle said, finding it hard to disguise the disappointment in her voice. "I'm going to have Buck give you a ride."

Before Michael could stop her, Isabelle yelled out the door, "Hey, Buck! The professor needs a ride!"

She turned back to Michael more rapidly than he'd expected, and it seemed that he was scrutinizing her, perhaps even admiring her. He reverted to a courteous neutrality when she caught his gaze, but for a brief moment, a flicker of unguarded appreciation, and perhaps even affection, flashed between them.

"You know, out there beyond the dome," Michael said, "I thought I saw something, too. Yes, it looked like a lizard."

They shared a second brief, unabashed glance before Buck shoved a helmet into Michael's hands and said, "Let's go."

Michael followed him out. As Isabelle closed the door behind them, a smile spread across her face. She allowed herself a moment to let the unexpected pleasure of recent events sink in before racing down the hall to her bedroom.

A large digital display was projected against the back wall of the room. Isabelle swiped her hand a few times across the screen and tapped a sequence of icons. A few seconds later Abe appeared on the screen.

"What is it?" he said.

"They're lying to us!"

"I know that. I've told you that."

"I need to borrow Lance."

"Because?'

"Give me forty-eight hours," Isabelle pleaded, a note of desperation in her voice. Then, as if to justify the request, she raised her left hand, in which she was holding Michael Costello's ID card.

10

Michael's motorcycle trip to the ECA building was swift but far from restful. He had sobered up enough to be distressed by the fact that his life seemed to be in danger at every bend in the highway. He had trouble remembering to lean into the turns as Buck had instructed him, rather than resisting them. But the news Alex had given him justified the hair-raising speed of the journey, and when he arrived safely at the facility he was thankful for the precious time Buck had saved him. He leapt off the motorcycle and handed Buck the helmet with an appreciative grin. Buck offered a tight-lipped smile in reply, as if to say, "I *guess* you're okay..." Then he sped off.

Alex was waiting in the lobby.

"Mike, this is bad. This is really bad."

"What happened? What's happening?" As they stepped into the elevator Alex said, "I think we hit the fast-forward button to the end of the world. I'll show you when we get to the lab."

A few minutes later Alex was at his station, his hands deftly opening menus and entering security codes that led them finally to the relevant screen. A few crew members were also at their desks. An atmosphere of strain, if not fear, pervaded the room.

When he arrived at the location he'd been looking for, Alex stopped.

"There," he said, pointing at something that looked like a large storm front. "That weather event is a lot larger, and a lot meaner, than anything we expected."

"Well, what's causing it?" Mike said.

"The first question that comes to my mind, Mike, is what *is* it? I'm pretty sure it's the flame front, significantly magnified by the methane explosions and the interactions with the jet stream. This is creating a much larger atmospheric disturbance than we intended, or are ready for. The data indicates that similar weather events are erupting in other parts of the world."

Deborah, who was sitting nearby, remarked dryly, "That's the understatement of the century." Michael watched what appeared to be lightning flashes on the screen, then looked up at the familiar orb in the center of the room, where thunderstorms of fire and rain were erupting intermittently across the earth's surface like kernels of corn exploding in a cosmic popcorn popper.

"So you're telling me that this particular flare-up isn't an anomaly, but the harbinger of further disasters?" Michael asked.

Deborah chimed in again. "The size is unique, so far. A thousand miles in diameter and it's generating winds of more than two hundred miles an hour. We couldn't survive too many like this one. And it's expanding. At its current speed it will reach us in fourteen hours."

"And it's gathering speed," Alex added, "which is consistent with the physics involved, if my etiological suppositions are correct. We've got to do something, Mike."

"Like what? Have you contacted the president?"

"He's not responding."

"What do you mean? That doesn't make sense. What do you mean he's not responding?"

"His office is stonewalling us," Alex said, shrugging his shoulders.

"What? Why?"

"Michael, we have no explanation for the president's behavior. Live with it. What we have is a firestorm, and I think we ought to do something about it if we can."

"Right now we simply don't have the authority," Michael said. "Which is probably just as well, because I'm not sure what we'd do in any case."

* * *

The office of President Erik Reyes had received numerous messages from the ECA lab, and he was keenly aware of the unfolding drama. Rather than respond to or take charge of the situation, however, he chose to focus his attention on a single aspect of it—the one that concerned him. At an enormous entrepôt a quarter-mile below the dome, a company of workers had been busy transferring supplies into a heavily fortified bunker. Hundreds of men hustled back and forth with forklifts laden with food staples, water, and other sundries. The men involved were familiar with the procedures—they'd been stocking the bunker for months. It was a high security operation, and the individuals involved had been chosen for their loyalty as well as their ability. It was only normal for the high-level authorities of an advanced nation to protect and provision themselves against emergency. But common citizens might well have been surprised and angered to learn how rapidly the program had accelerated in recent months, while their own rations were being reduced. In response to news of the approaching firestorm, the president was now calling for a ten-day around-the-clock operation that would increase supplies in the bunker far above the standard threshold.

He had just spoken with the lieutenant overseeing the operation, and he'd found it necessary to dress him down a little. The officer couldn't countenance the exaggerated scale of the new operation and had requested his confirmation code. It was a legitimate request, and Reyes should have been impressed with the officer's integrity and devotion to duty, but it frankly irritated him. And he'd just given the man another order to carry out. Complete the requested transfer, not in a matter of days, but in ten hours. He saw the lieutenant's jaw drop on the video screen before he snapped to attention and replied, "Yes, sir."

That unpleasant business having been attended to, President Reyes terminated the call and requested that

one of his agents in attendance see about transferring his personal belongings to the bunker.

The man nodded and beat a hasty retreat.

Then, to the one agent remaining in the room, the president said, "Bring her in."

* * *

Enormous storm clouds were building in the Mojave Desert to the east of the city. In some places the billowy masses stretched across the horizon, pulsing with a strange orange light. As the clouds approached, individuals began to notice the change in light, and then heard a very unusual sound: raindrops splattering across the surface of the dome.

Younger residents of the city had never heard that sound before. They opened doors and windows, or stepped out onto the street to look up at the rain as it smeared the dirt and grime high above them.

News reports of the Blue Skies operation had been

uncharacteristically vague, considering how much hype had accompanied the program's development and implementation, but now people began to surmise that it had been a great success. Wasn't this what was supposed to happen?

"They did it..." some whispered, then yelled. Strangers hugged one another on the street.

"They did it! They did it!" But it was a strangely muted and cautious celebration. The rain was real. But typically the government would be tooting its horn more loudly than anyone else if it had scored an important success. And this was the most important success of all. It topped screen doors and sliced bread by a long shot. Therefore, rain or not, it was hard to shake the thought that something had gone wrong.

* * *

Isabelle and Lance strolled through the lobby of the ECA lab as if they had worked there for years. At the electronic checkpoint Isabelle slid her wrist under the scanner, a flesh-colored wrist band covering her own tattooed wrist. The scanner beeped as it read the barcode, and Michael Costello's image appeared on the screen.

"Good afternoon, Dr. Michael Costello," the scanner intoned. "Please remain still for retinal scan."

Isabelle shot Lance a panicked look. He smiled cockily, pointed to the earrings he'd given her in the car before they arrived, and shook his head as if to say: "No worries."

As the retinal scanner focused on Isabelle's face, the earrings projected a holographic image of Michael in front of it. A moment later the entry light switched from red to green. "Thank you, Dr. Michael Costello. Have a nice day," it once again intoned.

At that moment Michael Costello was storming across

the same ECA lobby with Alex and Deborah right behind him.

"Be reasonable, Mike," Alex said. "You can't just storm the president's office."

"I can and I must," Michael replied. "What's the worst that can happen? Well, whatever it is, it won't be much worse than what will happen to everyone if we *don't* do something soon about the flame clouds. These are desperate times. You sent him—what?—sixteen messages over the past six hours."

Deborah pressed the elevator call button. Just then Michael turned and noticed Isabelle and Lance making their way across the lobby. Isabelle stopped briefly to make eye contact with Michael before Lance grabbed her by the arm and they hurried on out of sight.

"I wonder what *she's* up to?" Michael murmured to himself as he stepped into the elevator. He had the idea it might be something subversive … but also worthwhile.

Once the elevator doors had closed Deborah said, "What if he refuses see you?"

"That won't be one of the options. When we reach the reception area, just keep smiling."

Michael was a familiar face, if not a frequent guest, on the presidential floors of the building, and once he had reached the presidential offices, he strode toward the receptionist with an unhurried pace and a confident smile.

"Good morning, Dr. Costello. We weren't expecting you today," the receptionist said cordially. But before she could inquire as to his business, he picked up his pace and stormed past her desk without a word.

"Excuse me, sir … you can't go in there," she shouted lamely after him as she reflexively pushed the alarm button on the floor under her desk.

Costello yanked open the doors of the presidential office and entered with his two assistants trailing close behind him.

President Reyes was sitting behind the desk, taking a call on his cell. "I am not interested in your excuses. Find her. Now!" he said, before terminating the call and looking up in surprise. Meanwhile, one of his agents had made it clear with a single gesture that Costello would suffer brutish consequences if he took another step toward the president. Costello stood stock still. "We need to talk, now," he said in a quiet voice.

President Reyes made a familiar gesture as if to say to his guards, "No harm here, let him have his say." The men relaxed, and Michael sat back on his heels.

"My department has been trying to reach you for hours. I cannot stress enough the seriousness of this situation, Mr. President."

"Yes, I was notified of the situation last night, and we're taking proper actions."

"What the hell is that supposed to mean?"

"Calm down, Dr. Costello. We have a protocol for this type of scenario. Don't worry. You, your colleagues, and their families are all on the list. You'll all be safe."

"What is this list?"

"As our environment has deteriorated, we've realized the necessity of insuring that in the event of a catastrophic downturn in living conditions, some portion of the population could survive. We've constructed a bunker, if you will, and made sure it's well stocked with supplies to ride out just such an event as is now taking place. In about an hour we'll begin moving families in. We *will* survive this crisis."

"What?" Michael exclaimed, flabbergasted. "And how many families didn't make this list?"

"Let's just stay focused on the ones we can save."

"I didn't realize choosing which citizens live and which die was part of your job description," Michael spit out, making no effort to hide his scorn.

"Be careful, Costello. In order for us to save the human race, we'll need those who can help us rebuild it. Scientists. Engineers—"

"—politicians," Michael cut him short.

"Leaders, too, will be in short supply," President Reyes said calmly as he poured himself a stiff drink.

"And you're not even going to warn the common citizens of what's on the way?"

Reyes glanced over to the window, though he could see nothing but red sky from where he sat. "What would that accomplish? Nothing. Nothing but anarchy in the streets."

"But millions of people are going to die!"

"And fifty thousand will live! I'm doing my best to prevent the extinction of the human race. The other world domes are doing precisely the same. It's our responsibility to ensure that advanced life on this planet continues."

Michael said, "I have taken an oath to preserve the safety and advance the well-being of our citizens, yet you've spent the past ten years focused on how to save your own ass."

"As well as the families of all those who have dedicated their lives to protecting and serving this nation, including yours," the president said.

"You would be correct," he continued, "in your assessment of my efforts, if our last-ditch bunker project was the *only* one I'd sponsored. But I've also financed and supported a raft of other programs and projects. You were in charge of the most glamorous and expensive one—the one that everyone had pinned their hopes on, the one that seems to have failed miserably, stirring up more trouble than it alleviated, while further depleting our precious resources. Blue Skies was *your* project. Remember?"

Michael glared at President Reyes, but said nothing.

"Look, Michael, we have done our best to protect the citizenry. Could we have done more? Perhaps, though I don't see how. But whining about 'should have dones' and 'could have dones' isn't going to do us a damn bit of good during the next few hours. We have to take whatever steps are necessary to get us through this crisis."

"The crisis we created?"

"The crisis our forefathers left us with! Don't let personal feelings cloud your judgment."

"No, I won't," Michael replied. "Not anymore." Then he abruptly turned and stormed out of the room, followed by his two assistants.

11

The alarm bell of the digital display above the doors of the bunker screamed briefly, announcing that a new hour had been reached.

"We have four hours left," the foreman shouted into his microphone to no one in particular. "Everyone's doing well, but let's get a move on, people."

One of the workers took the opportunity to duck into a secluded corner of the warehouse and retrieve a tiny communication device from his pocket. "Four hours, and those doors are gonna lock," he whispered hurriedly once he'd dialed a number. "Whatever's going on, it's big."

In the distant hold of a casino barge, Abe sat on a crate of weapons as he spoke into his device.

"Are the explosives set?"

"Yeah. Everything is ready."

"Excellent. We'll be there at the appointed hour."

Abe hung up. He looked at the men sitting all around him.

"Change of plans," he said. "We strike now."

"But what about your sister?" one of his comrades said.

"We don't have time for that cloak-and-dagger crap," Abe replied dismissively.

*　*　*

Meanwhile, back at the ECA building, Isabelle and Lance had made their way into the mainframe terminal room. After a quick look to make sure the room was empty, they stepped inside, closed the door, and pushed a table up against it. Then they approached the heavy equipment.

"I don't think people come in here too often," Lance said, "unless they have problems with the servers. They can access what they need remotely through the network." He pulled up the holographic screen and set to work.

"How much time do you think you'll need?" Isabelle asked.

"I think this should go well," he replied. "I was expecting the firewall to be more difficult." His fingers danced across the screen as he penetrated deeper and deeper inside the mainframe.

"I'm in," he said finally, with a little spark of surprise.

With Isabelle looking over his shoulder Lance typed in "Project Blue Skies," and a series of files appeared. Isabelle immediately spotted a restricted file titled PROJECT: SLUMBER.

"There, that one. What's that?"

Lance went to work on it. <ACCESS DENIED>.

"Give me a second," he said. Before long a series of bunker blueprints materialized, along with technical specifications and a long string of unintelligible data.

"Bingo!" he exclaimed.

"Abe was right," Isabelle said.

"And take a look at this file," Lance said, opening another document. "It appears they've already drawn up an invitation list to this five-star resort."

"Hmm. I don't see my name anywhere," Isabelle said in mock disappointment.

"Mine either," Lance replied. "But your boyfriend is on it."

Isabelle stared at the screen: DR. MICHAEL COSTELLO. "Download it all."

"Why?"

"People have a right to know what's happening here."

At that moment a countdown popped up on the screen: 3 HOURS, 30 MINUTES.

"That's interesting. A system-wide notification. Something big is going to happen soon. I wonder what?"

Running her gaze down the documents in the folder, Isabelle pointed to one. Lance clicked and the screen came alive with forklifts in motion, sacks and bundles and water canisters being hauled on a busy loading dock.

"They're going to seal the bunker!" she said.

"Right now? I thought this was a 'last resort' type of thingy."

"Something's not right. Well, I guess I know what's not right. Blue Skies failed ..."

"What?"

"Michael told me the project exacerbated the condition—made it worse. But how much worse? Can you pull up the weather from here?"

Though he'd found it relatively easy to hack into the mainframe, it took a while for Lance to return to the common sites where weather data and forecasts were to be found. What they found did not look good. Radar showed massive storms rolling their way over the mountains to the east.

"I don't know weather," Isabelle said, "but I've never seen anything like that. We've got to get this out on the news feed. Right now. Can you do that?"

"Yeah, yeah. Of course I can. But what exactly are we

looking at here?"

While Lance set to work, Isabelle opened her communication device, hoping to ring up Abe. "Come on, Abe! Where are you?"

As she waited for an answer she glanced out the room's one small window and noticed a disturbance of some kind going on near the gate.

"I wonder what *that's* all about," she mused. The odd event only added to her growing sense of apprehension.

"I'm transferring the data now," Lance said. "It should be broadcast automatically in roughly fifteen minutes."

Lance returned to the umbrella folder where another set of files almost immediately caught his eye. Opening the subfolder, an array of file names appeared that sent red flags flying.

"This is something unusual, Isabelle. You'd better take a look at it."

"What is it?"

"The folder is named PROJECT ARK."

"Never heard of it."

"Well, it's heard of you," Lance replied. "Here's your photo." While Isabelle stared in disbelief, Lance made his way through a few more files, pausing briefly to examine text and photos.

"Very weird," he said. "You seem to be one of the stars of this operation. And look at this photo. That's Dr. Grace, the old zoologist from the San Diego Zoo. Sort of a folk hero to the retrogrades. Some admirers called him Noah. Quite Biblical. Anyway, he's identified in the caption as your father!"

"That's absurd. Abe and I were … they told me my parents were teachers, killed in the evacuation."

"And here's a video of you with some of your current

students. I wonder how they got that."

"And I wonder why they'd be *interested* in that," Isabelle said. "Surveillance has its uses, but this is going too far."

"Well, let's see. You've just broken into the government mainframe. You spend a lot of time distributing illegal medicines. Your brother is the head of a rebel militia. You're in love with the mastermind of the project that's going to save the *entire* world. Or not. Who *should* they be spying on, if not you?"

Suddenly there was a loud banging on the door. It opened slightly before striking the table in front of it with an even louder bang.

As agents struggled to push it farther open and enter the room, Isabelle opened the window and they both climbed out onto the fire escape. By the time the agents had freed the door and raced over to the window, Lance and Isabelle were several flights below.

"Stop them!" one of the agents shouted. "But don't kill them."

Another agent arrived at the open window and took a reckless potshot that ricocheted harmlessly off the metal staircase. He then climbed through clumsily and continued his pursuit, though his progress was hampered by the heavy weapon he was carrying.

Isabelle and Lance reached the end of the stairway and leapt to the ground. "This way," Isabelle said as she hurried toward the street, and then on to the disturbance she'd noticed a few minutes earlier through the window. "I think we can lose them in the midst of that ruckus."

12

At the moment Isabelle had called him, Abc wasn't answering his phone. He was sitting in the passenger seat of a garbage truck, fully decked out in old-fashioned military gear. The camouflage was sadly out of date, considering that green vegetation had largely disappeared from the region. (No wonder they'd gotten it so cheap

from the desert rats who sold it to them.) The truck was approaching the ECA building, and it almost flattened a distracted Michael Costello as he crossed the street to the parking lot following his interview with the president. The sight of a garbage truck was nothing extraordinary. But this was one of many such trucks moving throughout the city simultaneously as part of a clandestine operation that had been brewing for months.

Only slightly flummoxed by his near-accident, Michael answered his communication device as he reached the curb on the opposite side of the street. It was Alex.

"Mike! A swarm of men has taken over the computer rooms. They're confiscating all of our data, our research. What's going on?"

"First of all, don't do anything heroic," Michael said. "Let them take the stuff. It might turn out to be a good thing."

"What?"

"Did they tell you anything about the bunker?"

"Yeah, and they're moving me and my family into it. We're on some kind of list. Is this really happening?"

"I'm afraid so."

"You're coming too, right?"

"Don't worry about me. Just get Stephanie and Bree in safely."

Michael hung up and looked vacantly at pedestrians who were milling around in the streets and idly watching the rain splash against the wall of the dome. A sudden surge of dread washed over him. He spotted a group of children near the executive gates, standing with their hands and faces pressed against the wrought-iron bars. He glanced again at the sky, then at his watch. Then he headed for the gates, where a little girl was following a raindrop across the wall with her finger.

"Hey there. What's your name?"

The girl turned and stared at him for a moment before replying. "Amber."

"That's a beautiful name. Do you know who I am?" The girl nodded her head.

"I want to show you a secret," Michael said. Then he stood erect and began to walk backward toward the security gate.

"Keep watching," he said. Amber nodded.

Michael turned his head to the guard and said, "I'm Dr. Michael Costello. I'm going outside."

"Yes, sir." the guard replied, and held out a rebreather.

Declining the offering, Michael just smiled, walked to the gate, and slapped the exit button. A buzz could be heard as the air-secure gate closed around him and the gate to the outside world opened.

Amber continued to watch as Michael stepped out into the rain. He looked toward the sky and closed his eyes as the rain dribbled down his face. Other children had also begun to watch. Michael opened his eyes and smiled at Amber. He walked up to the dome wall and placed his hand against the glass directly opposite hers. Then he took a dramatically deep breath, and exhaled with equal force.

Amber's eyes grew wide. A little boy shouted, "He can breathe!"

Another said: "Look at him! Look!"

Passersby stopped and stared as Michael held his arms out and threw his head back, allowing the rain to wash over him.

"That's Dr. Costello," a woman said.

"What's he doing out there?" a man wearing a suit said. "No one should be out there."

"He's showing me a secret," Amber said.

Amber raced toward the gates, and a few of the other bystanders followed her. They pushed their way past the guard, who had little experience dealing with groups of people—few people ever left the dome for any reason. Soon an odd group of assorted men, women, and children were dancing together in the rain.

It made Michael laugh to see the uninhibited joy that surrounded him. Amber started to dance with an old man, children joined hands in a ring, couples kissed in the rain, a father tossed his young son.

But the joy was fleeting. Michael knew what the east winds were bringing. The rain was good, but soon things wouldn't be so pretty. Suddenly he saw something scurry out of the corner of his eye. Another lizard! He turned to follow it—and found himself face to face with Isabelle, who was panting heavily as the rain poured down her cheeks. Before he could say a word she blurted out, "Agents are on our trail. We were hacking into the mainframe, where we found all sorts of worrisome things."

"Calm down," he said, taking her by both arms and extending a perfunctory nod of greeting to Lance, whom he'd never set eyes on before.

"What's Project ARK?" she asked, looking him straight in the eye, searching for clues, for genuine honesty beyond the mysteries and double-talk of government policy and cant.

"How do you know about *that*?" he exclaimed, taken aback.

"I know the name. That's about all. But what is it?"

"I'll tell you what it *was*. It was hope."

"What do you mean 'was'? It seems to be a going concern, to judge from the files we've just seen." As she spoke

Isabelle once again scanned the crowd, but their pursuers seem to have given up the chase, or declined to join a happy and irrational mob in the hostile environment outside the dome.

"Well, then, I guess you know more than I do about it. The ARK was a repository for DNA. The blueprints for an enormous variety of plant and animal species were stored there by Dr. Grace—hence the Biblical name. No one knows what happened to the ARK, except perhaps for Dr. Grace's wife. And she's not talking. I spoke with her only yesterday—"

"She's still alive?" Isabelle said. Michael stared at her, wary and confused. *What did she know? How much did she know?*

Suddenly the government agents were upon them. They surrounded Isabelle and one of them jerked her arms roughly behind her back.

"You'll have to come with us," he said.

"What's this all about?" Michael said in a demanding tone.

"You are advised to stand back on orders of the president."

"Oh, the president. Why didn't you say so?" Michael said mock-reverentially, and took a step back before suddenly punching the agent in the face, knocking him backward onto the ground. A second agent trained his weapon on Michael's chest and said, "Stand back, Dr. Costello!"

"I wouldn't recommend using that thing out here," Michael said, though he did take a step back. "In this atmosphere it might electrocute us all—and in case you don't get it, that would include *you.*"

The agent looked at Michael warily, riled by his sarcastic tone yet not quite prepared to take a chance by blasting him. He held his finger against the trigger, then relaxed it, at which point Michael displayed lightning quickness in knocking him to the ground. Several agents grabbed him, turned him around, and began to pummel him with their fists. When they let him go he crumbled to the ground like a rag doll. One of the agents leaned over with a scanner and grabbed Michael's wrist. The scanner bleeped as it read the barcode.

"The president officially relieves you of your duties," the agent said, though it was doubtful Michael could hear anything that was being said. Ignoring Lance, the agents dragged Isabelle away.

Lance hurried over to revive Michael, who had a cut lip and also appeared to be spitting blood. He shushed a few curious bystanders away with a wave of his hand and a frown. Many had already fled once it became obvious that government agents were in the vicinity.

Michael came to almost immediately, and made an unsuccessful effort to rise. He'd blacked out and was now in a muddle, looking around for Isabelle.

"They took her," Lance said.

"Where?"

"I don't know where, and I don't know why. But those were some mean dudes. How you feeling?"

"I'll feel worse tomorrow."

Michael made another attempt to rise, and this time he was successful. He brushed some red dust off his clothing, and the two men headed back to the entry gate into the dome.

"I think all of this might have something to do with a project called ARK," Lance said. "You seem to know something about it."

Michael looked at him hard. "First, let me ask what

you know about."

"We came across some meaty files in a folder of that name. We hardly got a chance to look at them, but they definitely involve Dr. Grace—you may remember him. That 'Noah' character in the popular press. In one of the file photos he was identified as Isabelle's father. That's pretty bizarre."

"Her father? Isabelle's father?" Michael said. "I know exactly where they've taken her. I'll fill you in on the details on the way."

Back on the streets inside the city, Michael noticed a change in tone almost immediately. The arrival of rain had raised the noise level slightly, but now there seemed to be a note of hysteria in the air. As they hurried back to the ECA building Michael figured out why. The ubiquitous news monitors were displaying by-now-overly-familiar images that had been used to promote the Blue Skies project, but the message had changed. A computer-generated voice was repeating over and over again: "Project Blue Skies, the miracle we've been waiting for, has failed. Project Blue Skies, the miracle we've been waiting for, has failed ..."

Michael looked askance at Lance, who grinned sheepishly and said, "Yeah, about that. Isabelle wanted to get the news out. I don't now where she heard it. She sort of thought the people had a right to know."

"Of course she did."

The announcer continued: "The ECA would like you to believe that the rain is a sign of success. It's not. The ECA would like you to believe that the rain is a sign of success. It's not." The advertising logos for Blue Skies had now been replaced on the screen by images of workers ferrying supplies from a loading dock down a tunnel into an enclosed space that looked like a bunker.

The voice continued: "The rain is to distract us while the authorities race to save themselves. The rain is to distract us while the authorities race to save themselves."

Ensconced in his lofty office, President Reyes, too, heard these messages. "Can't someone turn that thing off?" he shouted. "I thought *we* ran the news. What's going on here? I don't care if we have to shut down the whole damn system! What difference would it make now?"

At that moment another agent arrived at the office.

"We have her," he said.

"Finally, something goes right," the president said, rising from his desk. "Let's go see what she can tell us." He sounded determined, but not very hopeful.

A half hour later the party arrived at the interrogation room of the prison. Laura sat calmly at the table, her hands folded.

"Should I be flattered? You don't usually visit so often," she said as they entered the room, in that mildly condescending tone that Reyes particularly hated.

"We're done with the games, Laura," he replied. He made a gesture with his head and two men entered with Isabelle in their grasp, her hands cuffed behind her back.

"You can't do this! I have rights!" she exclaimed.

"Shut up!" one of the guards said brusquely as he pushed her to her knees.

At this display of violence Laura jumped up instinctively, but she, too, was brutally forced back to a seated position.

"We don't have to do it like this," the president said. "The choice is entirely yours." Then he pulled a pistol from inside his jacket and held it to Isabelle's head. She screamed.

Ignoring the outburst, Reyes kept his eyes on Laura, whose preternatural confidence had given way to a distressed wariness.

"Once again, where did your husband hide the ARK?" From her kneeling position Isabelle looked across the room at Laura and slowly shook her head from side to side. Their eyes met.

At that moment Reyes smacked Isabelle across the side of the head with the pistol, sending her sprawling to the floor. Laura winced as Reyes strode over to the table and put his face inches from hers.

"You think you're being noble," he said. "You think you're preserving the world from an evil fate. But in fact, you're on the verge of destroying it. All the work you and your husband have done will be lost, and all of the life you've preserved will vanish. IS THAT WHAT YOU WANT?"

When Laura said nothing, Reyes continued: "You don't seem to understand. In a few hours this prison will be a watery grave and you'll be dead. I can move both of you to safety, but I need to know you're on my side."

"Your side? You want me to lead you to the ARK after the storm?"

"We'll need to rebuild. With the ARK we can create a new world."

"Do you want to rebuild, or control?"

Reyes's eyes narrowed. "Our goal will remain, as it always has been, to encourage lifeforms to flourish. We all benefit from that. I'm giving you and your daughter the opportunity of being involved in your husband's dream. Believe me, when this is catastrophe is over, I will leave no stone unturned until I find the material, with or without your help."

"It's a wide, wide world out there, Mr. President," Laura said. And then she shouted, "And how can you talk of letting life-forms flourish after smacking my daughter across the head with a pistol?"

"Think it over," President Reyes said as he and his men exited the room, slamming the heavy door behind them.

Laura rose from her chair and rushed to Isabelle's side.

"You're bleeding," she said as she helped the young woman to her feet. Isabelle was looking at her strangely.

"Is this true? Could you be my mother?"

Laura studied her face for a moment. Then her gaze fell to the dreamcatcher dangling from Isabelle's neck. She took a step forward and raised her arm, and Isabelle thought she was going to caress her face. But Laura reached instead for the pendant, which she examined intently. Then she gently let it drop and said, "Your father made that. He gave it to me on the day when our laboratory was raided. Later, after you were born, I pleaded with the hospital staff to allow you to keep it as a harmless memento. You know, Isabelle, the lizard is a symbol of healing and survival."

"I saw a lizard last night, just outside the dome."

Laura daubed Isabelle's wound gently with her shirttail, saying nothing. Then she said, "The elders believed that the lizard appears when it's time to look within and question where you are. They say he gives you strength to face changes and create a new life."

Isabelle said, "I believe it was a sign from God."

Laura smiled. "You may be right."

Isabelle said, "Project ARK. So that's what this is all about."

Laura nodded.

"What is it?"

Laura said, "It's ... everything. Everything that truly matters. Everything we as a people took for granted. It's the birds in the sky. The fish in the ocean. The beasts that once roamed the earth. Millions of species that once colored this world in a way no painting could. In a word, the ARK is life."

Laura stared at Isabelle.

"Are you going to help Reyes?" Isabelle asked.

"He may be the only person with the authority to implement a workable plan. Unfortunately for us, and for the world, I don't think he has an honest bone in his body..."

13

Soldiers, well-dressed VIPs with spouses and children in hand, and middle-class workers of every stripe stood patiently in line, waiting to be admitted to the bunker. Alex was among them, holding his daughter, Bree, tightly in his arms as he and Stephanie shuffled slowly toward the tunnel entrance. The clock above the door read 2:53. Alex noticed immediately that the numbers were descending and presumed the countdown marked the time remaining before the bunker would be sealed. It seemed a little hasty and that worried him. He looked around repeatedly, hoping to spot Michael. *What the hell is he up to?* he thought. *Always the hero.*

"I wanna go home," Bree said.

"Oh, no. This is going to be fun. Like a vacation!" her father reassured her.

"Really?" She sounded unconvinced, but her parents both nodded vigorously and then shared a brief happy-face look loaded with apprehension.

Suddenly a group of armed soldiers raced past them.

"That seems a little over the top," Alex said. But just then an announcement came over the loudspeakers. "Due to circumstances beyond our control, our timetable has changed. The bunker doors will be sealed within the hour!"

"Why is everybody running around like crazy?" Bree asked, on the verge of tears.

"Well, honey, a really big storm is coming."

"With rain and everything?"

"Everything. Even more."

"I've never seen rain before. Can there be something more than everything?"

"I think this will be something new to me, too," Alex said, trying to force a smile. A wave of fear washed across Stephanie's face, and Alex took her hand.

The general cacophony of the occasion was heightened a notch when a garbage truck rolled past. Then another one arrived. Soon it was a convoy of trucks rolling in.

Alex said nothing because he didn't want to disturb his daughter, but he was having a hard time fitting the pieces of this puzzle together. He could conceive of no practical use for so many garbage trucks in such a situation, unless the bunker had been supplied with tainted food, only recently discovered. And that wasn't a pleasant thought either.

They eventually reached the front of the line and Alex held out his wrist without making a sound. Green light. His wife and daughter did the same.

"All clear. You're free to enter," the attendant said. As they entered the bunker Alex took one last look out across the sea of privileged souls who would wait out the storm in the safety of that underground world. There was still no sign of Mike.

* * *

Out on the streets of the city, the news monitors continued to deliver a very unusual message. Scenes of underground stockpiling on an industrial scale alternated with graphic displays of a fast-moving storm approaching the city and footage of what appeared to be Melbourne's dome collapsing under the savage assault of fast-moving, superheated air and lightning.

"These are the preliminary results of Project Blue Skies," the monitor explained. "They're not encouraging. But where are the warnings for us? Citizens, find shelter wherever you can." The message then began to repeat. "But where are the warnings for us? Citizens, find shelter wherever you can..."

Then, suddenly, the screens went blank.

Michael and Lance had reached the steps of the ECA building. They turned to watch distressed pedestrians looking this way and that, as if they might find a suitable basement or bomb shelter near at hand. Others calmly went about their business, though many of them had picked up their pace. After all, the broadcast was very odd, and it made no mention of when the storm would arrive. Why would the government news service be casting aspersions on itself? And where did the footage from Australia come from? It all seemed like a devilish prank.

But if some found the televised warnings a little dire and hence hard to swallow, even the least gullible viewers soon became convinced, merely by looking around, that the streets would soon be full of chaos, and the dangers associated with such panic were very real indeed. People were rushing in all directions; Michael heard shrieks and saw a few elderly people being knocked off their feet. He and Lance were about to hurry on with their mission when Michael spotted a familiar face in the crowd. It was Buck, Isabelle's next-door neighbor. And he was coming their way.

"Buck," Michael called out and waved his hands. Buck spotted them and hurried over.

"Where's Isabelle?" Buck said to Lance, whom he seemed to know already.

"The Secret Service grabbed her."

"Maybe that's okay, for the time being. Keeps her out of harm's way. Abe called. They're on the move."

"What?" Lance replied. "Right now? The timing seems pretty bad to me."

"Yeah. But you know Abe. Anyway, he sent me to see if I could find you and Isabelle and pull you out before all hell breaks loose."

"It appears you're too late," Michael said, gesturing toward the mayhem that had reached a simmer on the streets just below them.

"That's not what I'm talking about," Buck replied cryptically. He scrutinized Michael for an instant—the bloodshot eyes, the streaks of sand on his pants. "Boy, it looks like you had a rough night."

"Too much celebrating," Michael said with a crooked grin.

"He doesn't know?" Buck said, turning back to Lance. Lance shook his head. Negative.

"Know what?" Michael said.

"I don't know how much you know about Isabelle, and you've probably never heard of her brother, Abe," Lance said. "But Abe has been unhappy with government control, resource allocation, that kind of thing, for quite a while now. He and his men have been planning a revolution."

"A revolution? You mean an armed insurrection?"

"Abe has never had great timing," Lance said.

"I wouldn't say that," Michael replied. "On the contrary, there could probably be no better time than during a period of widespread panic to launch a coup. All the same, unless he's got lots of help, including key men on the inside, it's not likely to succeed."

"Well, Buck is telling us that, for better or worse, Abe has mobilized a fleet of innocuous-looking garbage trucks throughout the city. They're being driven by dangerous men, many of whom have spent time on the outside. They're armed and they know a lot about the infrastructure of the city. Like you, they want to save the world. But they don't have your resources, and they don't share your confidence in the good intentions of the Federation establishment."

"Believe me, that confidence has been slipping away for a good long time, now," Michael said.

"I'm glad to hear it," Buck said. "Isabelle told me you were a good sort. Just a little naive."

"She told you that?" Michael stammered, unsure whether to be flattered or insulted.

Just then a deafening explosion rocked the steps and resounded off the surface of the dome. It was hard to tell which direction it had come from, but it was definitely somewhere nearby.

"That could be Abe now," Buck said.

"So what would be a sensible thing for us to do, in the context of these events?" Michael asked.

"What were you going to be doing?" Buck said.

"We were hoping to free Isabelle from the clutches of the government goon squad," Michael said.

"Mind if I join you?" Buck replied with a grin.

The three men raced inside the building and across the lobby to the elevators.

Just as they reached the entrance to the prison, a company of prison guards came racing out of the facility, heavily armed. They hid behind a dumpster, and Buck, who had considerable experience in clandestine operations, quickly laid out a plan for them. As the last few stragglers passed by, he stepped out from behind the container and leveled them both, rendering them both unconscious.

"Wow!" Lance said. Buck shrugged his shoulders as if to say, "That's what I do."

They hurriedly stripped the guards, and Michael and Lance put on their uniforms. "Hey, this is just like *The Wizard of Oz*," Lance said.

"Yeah," Buck replied, "but you've got to wonder where all those combat troops were headed. Abe is going to have his hands full trying to take control of the supply depot and the bunker."

As they talked Lance handcuffed Buck, and they continued down the hall with their "prisoner" in front of them. Michael had negotiated the labyrinth before, and it didn't take him long to locate the control room. Entering the large unoccupied space, they rushed over to the main console, and Michael started to explore the numerous surveillance feeds of various halls and detention cells while Lance clumsily removed Buck's handcuffs. Buck went back to the door while Michael and Lance continued to scour through the feeds.

Suddenly a large explosion rocked the building. "Things are getting nasty outside," Michael said. "We don't have much time. I had no idea there were so many prisoners."

"A man with nothing to live for has nothing to lose," Buck remarked sagely from across the room.

"See her?" Lance said.

"Not yet."

*   *   *

The crowds continued to swell, and the scene at the entrance to the bunker was deteriorating to a state not far from chaos, as the entry guards attempted to discern who had been authorized to be admitted and who had merely shown up in desperation hoping to con or force their way in. The arrival of a fleet of garbage trucks didn't help matters. It was likely to be the result of a program malfunction in scheduling, or possibly the belligerent attempt by a clandestine organization of employees to bully their way in to safety. But that would imply a degree of planning and lightning-quick response that maintenance engineers seldom exhibited on the job. In either case, it meant trouble, and the authorities had already summoned reinforcements. These reserves probably should have been on hand from the beginning, but President Reyes wanted to keep things calm and low-key. He also wanted to keep troops in reserve in case rioting broke out in other security-sensitive parts of the city.

Everyone had been caught off guard by the appearance of the nightmare weather system and also by the speed with which it was developing. And everyone would be caught off guard again as Abe climbed from his cab to the top of the garbage truck, a remote-control device in hand. He scanned the crowd below him haughtily, relishing the sight of VIPs ingloriously struggling to assert their prerogatives in the midst of the rabble they had been ignoring for years. Abe was not a poetic soul, but he recognized a moment of poetic justice when he saw it, and he smiled. Then his mind reverted to more practical matters, and, after scanning the crowd one more time, he keyed a few numbers into the remote.

A series of explosions resounded from the perimeter of the enclosure. At that moment Alex, who had passed through the checkpoint, grabbed Stephanie's hand, and they ducked inside the bunker wall. Bree screamed into her father's ear, "Daddy, I'm scared."

"Don't worry. I've got you."

Just outside, doors rose on the backs of the trucks as if on cue to reveal hordes of motley combatants inside. They had no interest in harming the common citizens in their midst but began immediately to search out anyone involved in government security. The task was made more difficult by the arrival of a new company of armed men who had no scruples about harming or killing an innocent civilian here and there. It was mayhem. Shots were fired. Some individuals fainted, some ran outside to escape the bloodshed, others pushed harder in their attempts to get inside the bunker now that the entry guards were engaged in combat with rebel forces. The sound of gunfire echoed loudly in the cavernous warehouse spaces, intermixed with terrified screaming. People were pushing in several directions like an angry surf, and anyone who ended up on the floor ran a serious risk of being trampled to death.

Deborah wasn't far behind Alex, and she made it into the bunker, where she almost immediately spotted him huddled with his family.

"Alex!" she shouted. "Where is everyone? Where's Mike?"

"I don't know. Haven't you seen him?"

Just then a massive earthquake shook the building—much more severe than the tremors that had rocked the city on a daily basis for years. Following the rumble there was a brief lull in the gunfire as combatants lost their footing or reached for the nearest support. Shelves of supplies came crashing to the floor, and well-stacked pallets of bulk commodities teetered dangerously. An overhead light fixture broke loose from the ceiling and dropped unceremoniously onto the heads of a senior executive and his pretty wife, rendering their bunker credentials useless. They would never move or breathe again.

Out beyond the walls of the dome, things were worse. The enormous tremors created deep and irregular fissures, running roughly parallel across the countryside, snapping antiquated gas lines from bygone eras and also breaking through to subterranean strata from which geysers of brackish water began to spout. As the hellish storm moved in, lose debris and the decrepit walls and roofs of abandoned structures near the dome were ripped away by the wind, to be sent airborne along with rivers of sand and grit. Some of the wind borne debris struck the surface of the dome with a deafening clatter. Lightning sizzled across the sky, igniting a few of the abandoned structures that held firm. The flames intertwined with orange and purple plumes of burning methane to spread devilish dancing arabesques across the sky. The self-regulating atmospheric and chemical effects that under normal conditions might have eased the ferocity of the reactions were rendered trivial by the snowball effects of the burgeoning firestorm. No one had seen anything remotely like it. It was as if the chemical reactions initiated by the Blue Skies project had unleashed a thousand years of seriously bad weather in the course of a few hours.

And the rains poured down. The water soaked into the parched landscape at first but soon began to rise and seep into the dome, which had been constructed to withstand the environmental attacks of a desert, not an ocean. The innocent joy and release of the first scattered raindrops had long since changed to panic, and chaos filled the streets.

On Church Row, the voice of Maxwell's digital

preacher struggled to rise above the surrounding chaos.

Now there was a strong wind, but the Lord was not in the wind. After the wind an earthquake, but the Lord was not in the earthquake. And after the earthquake a fire, but the Lord was not in the fire. And after the fire came a gentle whisper.

Maxwell sat beside his stand, listening to the wisdom of the scriptures. He'd heard it countless times before, but it had taken on new meaning, as he knew it would. Seemingly unconcerned by the commotion on the street all around him, Maxwell nodded sagely as the words re-sounded in his ears. Then he noticed a small boy in tears on the far side of the street. The lad had obviously been separated from his parents, and was lost, confused, and slightly terrified. Maxwell crossed over to him.

"Where's your mama, son?"

"I don't know. I can't find her," the boy said, trying to hold back the tears as he looked up into Maxwell's soiled but friendly face. Then Maxwell saw a group of children duck down the alley. Before long he had gathered a troop of frightened children under his care. He contin-ued to walk briskly through the streets, holding one child in his arms and another by the hand as several more fol-lowed behind. When he spotted a group of soldiers racing toward the back of the ECA building, where the loading docks and the garbage dumpsters were located, he knew which way to go. He yelled to some people on the street nearby, "Over there! That way!" Then he and the children raced off in the same direction.

* * *

President Reyes stood in front of the monitors with several of his personal agents looking on behind him. The events they were witnessing were beyond anything Reyes had imagined could happen. He had been under the impression, based on widespread surveillance and a broad network of informers, that elements opposed to the government had been largely eliminated, imprisoned, or driven out into the countryside.

"Who are these men?" he said to no one in particular. "And how could they have become so well armed?"

"Some of that weaponry looks out-of-date," one of his agents said. "And some of it looks like the armaments seized in that raid the other day."

Soon Lieutenant LeBlanc's face appeared on one of the monitors, and the president said, "Call in reinforcements if you have to. Under no circumstances are those rebels to reach the bunker! And prepare to seal the doors immediately. We'll be there within twenty minutes."

"Yes, sir. Copy that," the lieutenant replied curtly.

Turning away from the monitor, Reyes said, "We're moving Strongbow to the bunker."

"With all due respect, Mr. President, why are we still concerned with Project ARK, which seems to have reached a dead end, when the storm is nearly upon us?"

"Once all of this blows over," Reyes replied confidently, "that ARK will be the most valued currency on the planet. China. India. Russia. Australia. All of them our equals. But they'll all be clamoring to ingratiate themselves with us, because we have one very valuable thing they don't have—a path back to life. The bunker will save us today, but with the ARK, we control the future. Come on, let's go."

* * *

Once they'd determined which room Isabelle and Laura were being held in, Michael, Lance, and Buck sped off in that direction. The interrogation center was largely deserted, but when they reached the relevant door they encountered their old "friend" Hank, still on duty.

"Well, looky here," Hank said, "if it isn't the hero. Are you playing dress-up with your friends?"

"You need to let us in. It's a matter of life and death," Michael gasped.

"Everything here is a matter of life and death," Hank replied. "Are you referring to anyone in particular?"

"I'm referring to yours, mine, everyone's. You may not be aware of what's going on outside, but you must have noticed the earthquakes. And can you hear that roaring noise? That's just the start of a weather system that could blow the top right off this city."

Hank studied Michael's face. There was genuine desperation in it. Buck and Lance, on the other hand, both looked like they were about to smash his head into the wall.

"I wouldn't try anything foolish," Hank said. "I'm not as dumb as I look."

"Maybe not. But man, didn't you see the broadcasts?" Lance said. "We're not making it up. This shit is serious."

"No, I didn't see them," Hank said blandly, as if they were asking him about yesterday's baseball scores. He hesitated briefly, then said, "But I'm going to presume, judging by your uniforms, that you're officers of the government. I've never seen any of you before," he glared at Michael, "and I'm simply obeying orders. I have my reputation to uphold, after all. And I don't think you're going to find much in there that will help you save the world. If the top blows off this thing we're all cooked anyway, so what's the difference?"

He hesitated again momentarily, as if to enjoy his great moment of magnanimity, and then pressed the combination of buttons required to open the gate.

"Thank you!" Michael shouted over his shoulder as the unlikely trio raced down the hall.

Buck once again assumed the role of a prisoner as the little band passed Laura's cell, at which point the guards at the door pulled themselves to attention. They were not expecting the prisoner to deliver a blow that left them both unarmed and gasping for breath. A few additional jabs with a rifle butt from Lance and they were out cold. Lance then set to work on the lock.

Isabelle and Laura had heard the commotion. "It sounds like someone's trying to break in," Isabelle said, a touch of hope in her voice.

"That wouldn't be easy in this police state," Laura said grimly. Suddenly the door swung open and there stood Michael, Buck, and Lance.

The first thing Michael said was, "You stole my ID, didn't you?" Isabelle looked at him sheepishly and said nothing. But the look said everything.

"If I'd have reported it you'd have been toast the first time you tried to use it."

"I was counting on you not to," Isabelle said. Michael couldn't help grinning. At that point Isabelle jumped out of her chair and rushed into his arms. They shared a tender moment, oblivious to their surroundings, before Lance said, "Come on, Romeo. We gotta get out of here."

When Isabelle stepped back from his embrace Michael noticed the cut on her head. He touched it gently.

"Reyes," she said.

Michael winced and then said, "Let's go." Laura stood

and peered directly into Michael's eyes, as if she was looking for something, a secret sign of trust, perhaps.

"Yes, I believe it's time," she said.

Just then a voice from behind them said, "Dr. Costello. What the hell do you think you're doing here?"

Michael spun around to find Reyes and several of his agents standing in the doorway.

"Mr. President, you can't hold these women here. They'll drown. Water is already leaking in. We have to get them to safety!"

"These women are prisoners," Reyes replied. "They're involved in a Level 12—"

"I know all about the ARK," Michael cut him off.

Reyes' eyes narrowed as he studied Michael's face, perhaps looking for the same thing Laura had been seeking: a clue to where the man's loyalties lay.

"Then I guess you know this woman is a traitor to her country," he said, gesturing toward Laura.

"You're a traitor to the entire planet," Laura burst out.

Reyes took a step closer to her and said, "Have you considered my offer?"

"The Ark was never intended for one man to control," she replied derisively.

Reyes gave a quick, almost imperceptible nod to one of his agents, who deftly grabbed Isabelle and placed a weapon against her neck.

"Hey! What's going on here?" Michael said.

"Let her go!" Lance demanded.

Michael and Buck both pulled their weapons. Reyes' other bodyguard did the same.

"You're crossing the line, Costello," the president said. "You've given valuable service to the country and I've given you the benefit of the doubt, but you're going too far. I'm the president and I could have you liquidated in an instant."

"Tell him to let her go," Michael repeated. Reyes stepped closer—so close that Michael's weapon was touching his immaculate suit.

"You want to shoot me? Go ahead." The two men stared at one another, daring and menace in their eyes.

"Mr. President, I advise you to step back," one of the agents said. Just as Michael glanced his way, Reyes knocked the weapon out of his hand and then head-butted him. Michael stumbled backward. Buck took a step toward Isabelle, and a jolt of electricity surged through his body as the agent holding Isabelle fired. He collapsed to the ground, unconscious, and Reyes said, "We're out of time. Bring the women. And lock these traitors in."

One of the men held Lance and Michael at gunpoint while the other slapped handcuffs on Laura and Isabelle and dragged them out of the room.

"You can't just leave us here!" Michael shouted as the door slammed shut. Michael and Lance banged on the door as footsteps receded down the hall, sloshing through puddles of water as they went.

"You can't do this!" Lance cried. But of course they could. Michael looked down to see a trickle of water seeping in under the door. Buck was recovering slowly from the electric shock and he lifted himself up from the now-wet floor. He looked at Lance without saying a word. Lance looked up at the security camera above them.

"Let's get that," he said.

Buck rose to his feet, then hoisted Lance unsteadily, and Lance began to yank back and forth on the camera mount, being careful not to jostle the camera overmuch. It didn't take long for the lightweight carriage to bend and then snap.

"I've got it," Lance said.

Buck set him back down and Lance opened the back of the camera, exposing a tangle of wires. He examined them unhurriedly, nudging them with his index finger almost as if he were picking at a bowl of noodles, looking for the last piece of shrimp. His comrades looked on in reverent silence, but finally Michael said, "Can you do anything to help us get out of here?"

Lance deliberately dropped the camera on the floor, and it landed with a splash rather than a thud. "Not with this piece of junk."

"There has to be *something* we can do," Michael said.

"Yeah, pray," Buck replied.

It was an odd feeling, standing ankle deep in water in a locked cell, the water rising, and no means of escape.

"Do you think that table will hold all three of us?" Michael said.

"Probably not *you*," Buck replied. Then he chuckled.

"Maybe we should make a rope with our belts," Lance said.

"What would we do with a rope, hang ourselves?" Michael replied. "There aren't any windows."

"The force of the water might short-circuit the electronics in the lock, or force open the door by sheer water pressure," Lance mused as encouragingly as he could.

"Yeah, and monkeys might fly out of my butt," Buck said.

Suddenly a loud clang resounded from the door and seemed to echo loudly up and down the hall outside. Sloshing over to the door as fast as he could, Buck lifted the bar. The door opened! The water that had built up on the far side of it began streaming in. The three men fought their way through it out into the hallway, where they met up with a few other prisoners who were gleefully making their way to the nearest exit, whooping and hollering.

"Yes!" Lance exclaimed.

Buck smiled and said, "You see. Praying works."

Alarms were blaring. Lights flashed. The battery-powered emergency lights had come on. A mechanical voice was intoning over the intercom, "... to the nearest exit for evacuation. This is not a test."

Michael and his cohorts kept pace with the growing pack of other prisoners, some of whom looked pale and weak, as if they'd been behind bars for years. Some were probably hardened criminals. Others might have been freedom-loving souls who were overheard saying the wrong thing at the wrong time.

When they reached the elevators, Michael and Lance shared a look.

"Let the invalids take the elevators," Lance said. "Besides, they might stop working at any moment."

"This way. The roof!" Michael said as he dashed for the door to the nearby stairwell. Other prisoners had already found that route. On a landing they came upon Hank—who else?—who'd been flattened during the stampede, and perhaps roughed up by a passing convict with a grudge.

Helping him to his feet, Michael said, "What happened?"

"What happened? I tripped the emergency evacuation system, that's what happened. I couldn't leave all these people to drown!"

Hank joined them as they continued on to the top floor, where prisoners and a few of the remaining guards were heading for the rooftop exit.

"Reyes is headed for the bunker," Michael said. "He wouldn't have gone that way," and the three men ran in the opposite direction down the hall.

"You can't get out that way!" Hank shouted, before he, too, climbed through a window onto the roof of the building.

As they raced along the corridor, Michael, Lance, and Buck got their first look through the interior windows at the chaos below.

"Bloody hell," Buck said.

"There's Laura and Isabelle, out on the catwalk with Reyes," Michael shouted. "I think we can catch them." The trio climbed one after another out the window and shimmied over to a nearby fire escape ladder.

* * *

President Reyes and his men hurried along the catwalk that stretched high above the loading dock, virtually dragging Laura and Isabelle along with them. Beneath them chaos reigned, as rebels fought Federation forces while the elite among the citizenry struggled to make their way through the gates and into the bunker. Reyes knew he was taking a chance; if one of the rebels happened to see him, he'd be a sitting duck. But all the other avenues of approach to the bunker were more dangerous still.

The mayhem below the little party increased when a new throng of citizens appeared at the entrance to the loading dock, many of them children. Gunfire ceased momentarily as everyone apprised the new situation. Reyes turned to one of his agents and said, "Stop them! Those people are not to reach the bunker."

The agent delivered some orders through his headset. "Form a line. Hold them back." But he wondered how long it would be before survival instinct took over and the killing became more indiscriminate.

"Look what you've created, Erik," Laura said.

"I wish you'd drop that sanctimonious tone," Reyes snapped back. "As if you've done anything to help your fellow man in the last twenty years. I didn't cause this riot. And I didn't create the storm. We entrusted our finest minds to the task of returning the planet to health with no help from you, and they failed miserably. Maybe the science wasn't good enough. Maybe those who were administering the program lost their nerve. There's really not much point evaluating it now because we'll never have the resources to attempt anything on that scale again. One thing I did do was build a bunker to save lives and ensure that a rebuilding process could begin."

"Do you really think the lives on your precious list matter more than any of those people down there below us?"

"Yes, I know," Reyes said, as the party hurried toward the stairs alongside the bunker entrance, "that from a certain point of view no individual is better or worse than any other, and don't forget, not a single sparrow drops to the ground without the good Lord seeing it fall. Meanwhile, our slim hopes for the future rest with a cadre of talented people who know how to get things to work. That's just the way it is."

Suddenly an explosion rocked the building, sending everyone on the catwalk into the railing. As she struggled to regain her footing Isabelle spotted Cameron, one of her colleagues on the casino barge, lying bloodied and motionless in a heap below her, with Abe by his side, trying to bring him around. Then she saw the soldier coming into range behind them.

"Abe!" she shouted at the top of her lungs. Abe ducked

instinctively, spun around, and leveled his assailant with a single shot.

President Reyes could see the mob was gaining the upper hand by sheer force of numbers, and he instructed his agents to hurry ahead, lock down the area, prepare to seal the doors, and start raising the ramp.

As the ramp started moving the screams and cries intensified.

Within the safety of the bunker, Stephanie buried her face in Alex's shoulder, frightened by the wailing.

"Why is Mommy crying?" Bree said.

"She's sad for the people who can't get in here with us," Alex replied.

"Why can't they come in? I can move over."

"There isn't enough room, sweetie. But I can move over too."

A few seconds later Alex handed Bree to his wife, stood up, and headed toward the entrance to the bunker.

"What are you doing?" Stephanie yelled, an element of hysteria in her voice.

"I've got an idea," Alex said, shooting her a reassuring smile.

As he approached the entrance Alex ran into Lieutenant LeBlanc, as he'd expected he would.

"What are you up to?" LeBlanc said.

"I'll show you." And with that Alex walked over to the man operating the ramp, whom he knew slightly through associations at the lab, and said, "Let them in. There's plenty of room, and we can make more."

"I'm not risking my life to save that rabble," the man said.

Grabbing the pistol from LeBlanc's holster, Alex said, "You'll be risking your life if you don't."

"Alex, be reasonable..." the man said, as if he were

talking to a child.

LeBlanc stepped up alongside Alex, a second weapon in hand.

"You heard him ... Those people are coming in here."

More impressed by an unknown officer in uniform, the man stepped aside, his arms raised.

As he lowered the ramp, LeBlanc whispered to Alex, "Your safety's on." Alex looked at the pistol, but he had no idea where the safety was.

"Probably best we leave it that way," LeBlanc said.

"Yeah," Alex nodded.

Even before the ramp had been lowered completely, people had begun to lift themselves up and continue down toward the bunker. When it had finally dropped back into place, civilians poured across it. Meanwhile, the president and his reluctant entourage were approaching the far end of the catwalk when a loud voice behind them cried, "Reyes!"

The president turned just in time to receive a serious blow to the stomach, arriving out of nowhere, which flattened him. Michael had latched on to a stray pipe during his scamper across the catwalk. He now tossed it aside and grabbed Reyes' key remote from his belt to unfasten the cuffs on Laura and Isabelle. As Michael fumbled to help the women free their hands, Reyes pulled himself up to his feet and drew his weapon.

"What do you think you're doing, Costello? Drop the dimestore heroics and help us get to the bunker."

Just then another tremor shook the building. The catwalk dropped about a foot but held firm to its anchors. Reyes fired. Michael was riven by a electric surge, and he collapsed. But as the catwalk continued to shake, the railing panel that Reyes was leaning against for support broke loose. He spun around instinctively to grab it and ended up stretched out into space with his feet on the floor of the catwalk and his hands gripping the half-detached railing. Laura and Isabelle looked on in horror as Michael, hardly able to see or move, groped for Reyes' feet.

"Hang on," he shouted. But Reyes could not hang on, and as his fingers slipped his body made a swan dive into the crowd below him.

Still dazed, Michael looked up to see a small lizard directly in from of him. He stared, and the lizard stared back. *I'm hallucinating*, he thought. He closed his eyes again. His head ached.

"Michael!" Someone was calling his name, as if from far away. He opened his eyes again. Isabelle's dreamcatcher necklace was dangling in front of his face, the lizard prominent as it spun.

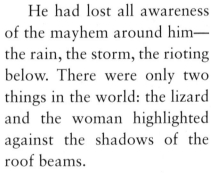

He had lost all awareness of the mayhem around him—the rain, the storm, the rioting below. There were only two things in the world: the lizard and the woman highlighted against the shadows of the roof beams.

"What's this?" he said, grabbing hold of the talisman.

"It's a dreamcatcher. My father made it. Come on, now. Get up—"

"Your father made this?" he exclaimed.

"Come on, now. Get up, Michael. We don't have much time!"

Oblivious to her entreaties, Michael returned his attention to the necklace. He studied it as if it were the only thing in the world. Dr. Grace had given this to Isabelle. He'd made it himself. Suddenly more attentive, Michael noticed a series of markings around the edges of the circular pendant, like spokes on a wheel. He pushed the lizard gently. It slid to the side, and the outer wheel began to spin. Michael looked over to Laura, who had stopped trying to wrestle him to his feet. Their eyes met.

"It's hidden in the stars," he said.

"What?"

Michael seemed to be regaining some mental acuity. "This is some sort of navigational device. An astrolabe," he said. "See these knots. They're stars. This is a celestial map!"

Michael rose to his feet, stumbled, but then moved quickly toward the stairs, which were now nearby. Isabelle was right behind him.

"I have to go," he said.

"What are you talking about? Where are you going to go?"

"You were right about the lizard."

"The lizard? Michael—talk sense."

"It's the ARK, Isabelle ... this is the map. I have to find it."

"Well, I'm coming with you."

"No. That's crazy. Be reasonable. The bunker is right there. Help your mother, and save yourself. Buck and Lance will go with you."

"I'm coming with you. I saw the lizard first, and it's my necklace."

"And so am I," Laura chimed in.

"No way."

"I think I'd be a good man to have aboard," Buck said.

"Aboard what?" Lance said. "Once Michael settles down, he'll realize he has no idea what the symbols on the astrolabe mean, and no means of transport if he did happen to decode the message. There's a massive storm blowing outside more ferocious than anything the earth has seen in a billion years. There's only one way any of us could survive for long outside that bunker."

"The casino—" Michael blurted out.

"You're not serious, Michael," Isabelle gasped. "There's no way to know if it will even float."

"I guess we'll soon find out."

Isabelle looked at Lance. "No one knows that engineering system like you do, Lance. What do you think?"

"I think it's suicide."

"Don't worry. I can run it," Buck said.

"What? You think *you* can run it?" Lance said, incredulous.

"Yeah, I've seen *you* do it."

"Watching and doing are not the same thing, man."

"Lance, don't worry," Buck repeated. "I got it."

"No, you don't got it. But I'm in. Life in a bunker never appealed to me much anyway."

"Then we've got to get out on the roof and grab a tram," Michael said.

"You think the trams are still running?" Lance laughed.

"If not, then we can make them run. This way," Michael said. And they raced as gingerly as they could back across the now severely damaged catwalk.

Once outside, they could see the dreadful red tumult above them. Clouds of sand and debris swirled by, and torrential rain pounded the surface of the dome above the ECA flight deck. They soon reached the nearby sky-tram terminal and climbed aboard the tram car at the front of the line. No one was around, though they could see a few people in the distance—disoriented ex-prisoners perhaps, unsure where to go under the unusual circumstances.

Lance found a control panel, opened it, and after a few minutes succeeded in getting the vehicle into motion.

"Hang on tight," he said. "This could be a wild ride."

The others grabbed hold of the overhead bars as the tram started to putter forward.

Looking out the window, Isabelle gasped. Michael crossed the car to her side, where they could see a section of the dome splitting open. Rain and hail began to pour down onto the streets nearby and noxious fumes swirled ominously.

"Lance, can you speed this thing up? We're getting close to lights-out here," Michael said.

"I'm at full speed! And derailing the thing wouldn't be a good idea. But the last section is a straightaway. We'll be fine…just so long as I can figure out how to stop it."

Michael could tell, as the tram car approached the casino station, that the stop would be brutal. He wasn't altogether sure they'd survive it.

"Brace yourselves!" Lance shouted as he applied the brake with all his might. The tram lurched to a stop, but less violently than Michael had feared. Everyone had been gripping an overhead loop, and though both Laura and Isabelle lost their footing momentarily, they hung on and neither was hurt.

"Ladies and gentlemen, thank you for traveling sky-tram," Lance said in an industrial voice.

"Quick. Go. Go. Go." Michael said as he leapt from the car.

They raced across the platform as sheets of water

crashed down. Hand in hand, Michael and Isabelle sprinted toward the casino barge as fast as they could go. Then they heard a voice.

"Hey, hero. Hey! Over here." It was Hank, standing beneath an overhang. He waved them over. "What are you doing here, you crazy son of a bitch? Headed for the barge?"

"Yeah. How did you know?"

"That's where everyone around here is going. It's the only thing inside this part of the city that floats. Come on. But it's flooding down on the street. We'd better take the bridge."

*   *   *

President Reyes made a soft landing on the floor of the ECA warehouse, courtesy of the thigh-deep water and several individuals who were knocked unconscious when he landed on top of them. Reyes' shoulder was severely bruised but not dislocated, and he got to his feet quickly. It occurred to him that the men he'd landed on might drown, but there was nothing he could do about it, short of dragging them to higher ground, one after the other. He didn't have time for that, and what would be the point? He began to push his way through the crowd toward the bunker as well as he could, just one among many desperate people. At one point he looked back to see a wave of water rolling down a corridor toward the warehouse floor. He was fifty feet from the bunker. Soon he'd spot some government agents who'd recognize him and give him an escort. They needed him inside.

One person had already recognized the president. It was Abe. He was heading in the president's direction when an alarm sounded and the bunker's giant doors began to close.

"WAIT! WAIT!" the president shouted. "That's an

order!"

Abe also started shouting "Hold the doors!" And they weren't the only ones shouting.

Inside the bunker, Alex held Bree close.

"Daddy, I wanna go home," she said. Alex pulled her in tighter. He felt someone looking at him, and looked up to see Maxwell, the preacher.

"I believe it's time to pray," Maxwell said in solemn tones.

Alex shook his head. "It wouldn't be convincing. You see, I don't believe in God."

Maxwell just smiled. "That don't mean He doesn't believe in you."

Just then a roar resounded through the bunker. It was coming from outside, where the vast wave had reached the loading dock and was wiping out everything in its path. Both Reyes and Abe were less than twenty feet from the bunker door, which was closing faster now. It was slow going through the standing water, like a bad dream in which progress becomes impossible. The door was closing. Not possible. A few ahead of them got in. Six feet...two feet...Then it clanged shut! Nowhere else to go. Water approaching. Must pound on the door. The roaring behind them.

Abe looked at Reyes. Together they turned, fear in their eyes, just in time to see the wall of water descend that would crush them against the wall of the bunker.

Inside the bunker, the arrival of the tidal wave made a profound and fearful thud, after which the lights blinked and went out. People screamed, which only increased the general pandemonium within the enclosure. Alex pulled a flashlight from his pocket.

"There's got to be a generator," he said.

"I know where it is," LeBlanc said as he rose to his feet.

"I'll come too," Maxwell said. "I'm good with machines."

Alex rather doubted Maxwell's engineering expertise but saw no harm in letting him tag along.

"That'd be great," he said, forcing a smile. LeBlanc led them back through the dark and now overcrowded hall.

*   *   *

Michael, Isabelle, and the others raced across the bridge that led to the rooftop of the adjoining building. Tables and chairs were scattered around, upended by the winds that had entered through the fractured dome. Buck deftly hurtled an iron chair that came rolling past him in the gale force winds. It might even have been fun if there hadn't been so much at stake.

"Here! Down this way," Lance said as he hurried down an escalator that had been paralyzed by the power outage. The group reached the bottom of the escalator and found themselves in the lobby of a glitzy office suite. Michael and Buck tried to open the glass doors, but they lacked the force to overcome the water pressure holding them shut on the other side.

"I can't believe the water has risen this high," Michael said. Buck grabbed a fire extinguisher from behind a reception desk.

"Stand back," he said, as he banged the metal canister against the sturdy glass. "When it breaks, hold your breath, grab on to something, and watch out for broken glass!"

After a few strikes the glass cracked and the water began to flood in.

Once the surge had subsided, Michael helped Laura through the doors. Others followed, and everyone headed for the surface. Up on the street the water was only waist deep. Laura wrapped her arms around a lamp post and scanned the group.

"Where's Isabelle?" she cried, but just then Isabelle's head emerged above the dirty and roiling flood.

"I'm here... I'm okay."

"There it is!" Hank cried as he spotted the Jolly Dolphin, directly in front of them, sitting slightly askance as water rose around it. People were climbing aboard in ones and twos.

As the group sloshed toward the gangway, Michael saw what looked to be a desert rat grab hold of the little girl just ahead of him. It was Amber, the girl who had watched Michael test the air outside the dome a few hours earlier. It seemed like a lifetime ago!

"Hey! Let her go!" Michael shouted. But as he spoke the rat lifted the girl up to safety and was pulled underwater by the effort. Michael dove in after him. A few seconds later he resurfaced. Buck grabbed for him but missed.

"Here! Give me your hand!" Buck said. The rat had also burst from under the water, gasping and thrashing about. He grabbed hold of Buck's hand while Lance struggled to get Michael up onto the gangway. When they were all out of the water Michael looked at the rat.

"You saved her—the kid," Michael said.

"Then I guess we're even," said a distinctly feminine voice.

Michael stared in shock as the rat hoisted herself onto the barge without difficulty. She was obviously in very good shape.

Just then the vessel shifted as the rising water lifted the hull, which was still fastened to the now submerged dock.

"We've got to cut it loose," Michael said. "I hope it isn't fastened in more than one place." Hastening down the deck to where the barge was tied, Michael and Buck took the plunge. A minute later they reappeared.

"Too tight," Buck said as he gasped for air. "Can't get it loose. Who's got a knife?"

Without a word Lance produced a pocket knife. "Why didn't you ask sooner?" he said as he handed it over.

Michael looked at it. It wasn't much, but it would probably work.

"Thanks," he said. "Now maybe you should escort the ladies inside?" Lance nodded and gestured gallantly to Isabelle and Laura to accompany him across the gangway.

After getting the rope to fray with the knife, Michael and Buck made a few more dives, and they finally succeeded in loosening it. As the surging water rocked the boat there were moments when the rope went slack, and during one of these they succeeded in unlooping it from the post. Keeping a hold on the rope they pulled themselves back to the vessel and struggled to climb aboard as the barge began to drift down the city streets. Hank and one of the staff of the casino, a man they called One-Eye, stood on board, pulling on the other end of the rope to help them up. Michael momentarily lost his grip; Buck lunged for him and then he too lost the rope. Michael reached the deck and hoisted himself up with One-Eye's help, but a moment later the drifting barge slammed into a building. Buck was down there somewhere, but it would be suicide to take a plunge in that suddenly constricted space.

"No! Buck!" Michael called as Hank pulled him back from the edge. They both watched the surface of the water in disbelief. Nothing.

"He's gone," Hank said finally, lowering his head. "He's gone."

Soaked to the bone, grief-stricken and exhausted, Michael, Hank, and One-Eye wandered below deck. The sounds of breaking glass and shouting people could still be heard from time to time. The barge was proving seaworthy, at least for the time being, but it was anybody's guess how long it would hold together as it was tossed like a giant cork against lamp posts and building facades.

"We've got to get this thing moving," Michael said. "We won't be able to steer it clear of trouble unless there's some power behind it."

"I'm pretty sure there's power in the storage batteries," Lance said. "That's what we used to run the casino—but the engine wasn't needed in dry dock, obviously, and it isn't on the grid."

"Well, how do we get it on the grid?"

"The quickest way would be to unplug something else. Like the slot machines. Then we could run that cable to the other side of the deck and I could hook it into the engine room. Trying to locate and restore the internal connections would take much longer."

"You mean on the deck? Out *there*?" Hank said.

"I'll do it," Michael said.

"Not alone you won't," Hank said, suddenly emboldened. One-Eye also agreed to join the party, and they emerged together out into the fierce winds and rain, fighting for every step.

"Lance said third from the right," Hank reminded them when they'd reached the cables.

"This is it," said One-Eye, putting his big mitts around the plug, shielding it from the rain with his body, and giving it a mighty tug. Once it was detached, the three men dragged the heavy cable across the deck and back inside the cabin.

Meanwhile, Lance had located a laptop and opened it

on the desk, which he'd cleared with a single mighty swipe of his arm. *I've always wanted to do that*, he thought to himself as he plugged the device into the system. It was slow to boot up and Lance muttered "Come on, come on, COME ON!"

"System operational: Welcome," finally appeared on the screen.

"Ah, yes! You're beautiful."

Leaving the desk he sprinted out of the room and down the hall. The plate on the door he opened said ENGINE ROOM. Inside were stacks of supplies and assorted junk, behind all of which was the engine panel, mounted to the wall. Lance scanned the controls, then flipped a switch... Nothing.

"Talk to me, baby," he said, looking around to see where the trouble might lie. He finally spotted a cable wrapped in a coil. Pulling another box aside, he knelt to see if he could figure out where to reconnect it.

Meanwhile, Michael, Hank, and One-Eye had broken into the pilot house. It was dry inside, though the walls were rattling fiercely in the wind. Michael wiped a thick layer of dust off the navigation panel.

"It's ancient," Hank said. "Any idea how it works?"

"Well, this looks like ON," Michael said as he flipped a switch. Nothing happened. Hank and One-Eye looked at each other and shrugged, suppressing their smiles.

But in the engine room the control panel suddenly came to life, whirring and blinking. Lance's eyes, too, lit up. Pressing the intercom button, he said: "Michael?" A moment later a voice replied, "I'm here. Fire her up."

"I've been trying," Lance said. "I think we're getting closer. Hang on." He tried a few more settings. Still nothing. From the pilot house Michael said, a note of urgency in his voice, "We're about to crash head-on into the Wells Fargo building. I'd love to throw this thing into reverse… real soon…Any time now, Lance."

At that moment Isabelle burst into the pilot house. "We're heading straight for that building!"

"I see that," Michael said drolly.

"I'm working on it!" Lance's disembodied voice once again filled the little room. He tried another combination of settings, then began turning things on and off at random, and finally gave the panel a swift kick.

It sprang to life, blinking and purring.

"And…we have power," he said, exultant. Up in the pilot house Michael gingerly nudged the controls, which he'd been studying carefully. The barge turned ever so slowly. Barely avoiding direct impact with the building, it began scraping along the side, making a horrible grinding noise as it scoured the facade, freed bricks from windowsills, and snapped off wrought-iron decorations.

Inside the casino, the scraping noise was muffled but clearly audible—just one more bizarre and unexpected horror of unknown origin to add to the list. The passengers—former inmates, guards, local residents, and longtime bar patrons—mostly sat in stupefied silence, happy to be out of harm's way for the time being but incapable of mustering the energy to do anything constructive. Anyway, what on earth could they do?

Some individuals huddled together in small groups. Others sat in stony silence. A few seemed to be praying. Amber sat in a corner, alone and scared. Several wary eyes looked on as the rat who had saved her earlier crossed the room to sit beside her. The little girl seemed relieved when the rat took her hand, and so did the onlookers.

Outside, the weather remained fierce, and Michael did

his best to maintain some speed, which would make maneuvering easier, though he didn't have anywhere in particular to go. Massive waves appeared at irregular intervals, no doubt generated by buildings that had crumbled in other parts of the neighborhood. Michael's plan was to move the barge toward the wall of the dome, where buildings were shorter and less likely to fall on top of them, but the wisdom of that plan was put to the test when a large wave appeared out of nowhere from directly behind the barge, picked up its stern, and sent the entire vessel roaring toward the wall itself like a circus ride.

"Brace yourselves!" Michael shouted into the intercom. "This could be serious!"

Everyone in the pilot house wedged themselves into a corner or grabbed a fixture or railing with both hands. Michael gripped the edges of the control panel, wishing that the barge had an old-fashioned wheel to hang on to. The impetus of the surge under the weight of the barge was such that, like a surfer catching a perfect wave, it propelled the craft for a considerable distance. By the time the wave dissipated, the barge had picked up speed, and it was obvious to Michael that it was going to ram directly into the wall of the dome itself. He threw the engine into reverse and attempted to turn the vessel, but the engine was a weak tool in the face of the momentum, and the barge crashed headlong through what remained of the dome wall. Everyone was thrown to the floor or into the nearest obstacle. Michael struggled to regain his footing, then looked around hurriedly to see if anyone else had been injured. Hank had clunked his head, and a few elbows had been bruised, but no one seemed to be seriously hurt. Turning his attention once again to the task at hand—navigating the vessel—Michael was taken aback by the strangeness of the scene that confronted him. They were outside the city now, bobbing on immense swells. There were hills in the distance and mountains behind them but they looked altogether different in this strange light and surrounded by water. Outside the protection of the dome the winds were more fierce and the lightning more dramatic and frequent. The seething water seemed to be on fire in some places. There was nothing much to run into, which was good, and in the absence of obstacles to upset its flow, the movement of water across the landscape was more powerful but also slightly more predictable. Though Michael's primary concern was to handle the ever-shifting waters in such a way as to keep the barge upright, he was also trying to maintain his bearings with respect to the local landmarks. If they succeeded in surviving the deluge that was now upon them, their search for Dr. Grace's remote laboratory, under Laura's guidance, would begin in earnest, and it would be important to know as precisely as possible where they were.

Night was approaching. Michael looked back out the pilot house window to see the city, like an antique oval greenhouse, shattered and undone, but still glowing with flashes of eerie light. But it was much smaller than Michael had expected—they must be moving at tremendous speed! And the shallow arch of the distant structure, which could be discerned even though much of the surface material had been shorn away, suggested that only the very top of the dome remained unsubmerged. The volume of water involved couldn't be the result of rain alone. Fishing for explanations, Michael wondered if it had been released from reservoirs deep within the mantle of the planet as a result of rapidly shifting tectonic plates. If that was the case, then it was also possible that a tsunami of astonishing proportions could appear unexpectedly from the west at any time.

Under other circumstances such thoughts might have triggered a surge of fear, but Michael had been through so much in the last few days, and continued to bear such a burden of guilt due to the ongoing catastrophe sparked by the Blue Skies project, that a mere tidal wave hardly registered as something to be dreaded. The barge, the crew, Isabelle and Laura, the assortment of passengers they were ferrying eastward into the unknown across the half-submerged Mojave Desert—that was his world, and he chuckled a little as it occurred to him that they might actually be passing above Death Valley at that very moment. Would the air soon become unbreathable or the heat too intense? There was no use speculating. And what was their destination? Michael had no idea, though he was eager to find out. And that thought brought his attention to bear once again on Laura. Did she know where Dr. Grace's cache of genetic material was? Could she find it again under these radically different conditions? If so, would it be of any use to them?

Hank had recovered from his knock on the head and come forward to stand beside Michael, who was still deep in thought. As they looked out across the waters together, Hank said cheerfully, "They don't build 'em like this anymore," and he patted the side of the instrument panel gently.

"They don't build them at all anymore," Michael replied stonily. But then he realized that Hank had been speaking facetiously, and he couldn't help cracking a smile. Brooding wasn't going to do him or anyone else much good.

Isabelle noticed that Hank's remark had lifted Michael's dour mood, and she, too, came up to the window. He turned to look at her, and for a split second his mind was filled with a single radiant thought: *just getting to know you has made this entire fiasco worthwhile—for me.* It was a selfish thought but it felt good to think it, and Michael experienced a vital surge throughout his body.

Without saying a word, Isabelle handed him the dreamcatcher. He looked at it. Then, puzzled, he looked at Isabelle. He looked at the dreamcatcher again, with the lizard in the middle like a cameo in a brooch. He touched the lizard, nudged it, and found that it slid to the side; as it did so, the outer wheel began to spin slightly.

Hank was watching the proceedings over Isabelle's shoulder, and when he saw the spinning dial he said, "Hey, that might be an astrolabe of some sort. Let me see that."

Michael handed him the object, and Hank took a closer look at the dial. "Yeah... had one of 'em when I was a kid. It was just a science project, but I really got into it. Old-fashioned technology. It didn't further my career a whole lot. We got somewhere to go?"

"We do," Michael said. "Do you think this device will help us get there?"

"All by itself? Of course not. We'll need two things. A set of coordinates to identify our destination, and an occasional look at the night sky."

"Hank, if I had coordinates, I could just use the GPS," Michael laughed.

"I'll tell you something, doctor. The mega-solar surge that powered the Blue Skies project also took a big bite out of the satellites that give us our 'global position.' That system is now shaky at best. But if we don't know where we're going, it's irrelevant in any case."

"Let me see that again," Isabelle said. "My father gave

this to my mother, and she gave it to me. It's true, she's Native American, but she's never been much interested in ornamental trinkets. Now we discover that this is a navigational instrument. I wonder what other secrets it harbors."

"It's a sophisticated piece of engineering, I'll tell you that," Hank said. "Aside from the very delicate incremental markings, it wasn't easy to design that lizard to hold firm inside the brooch through years of wear, but then start sliding when you really put some pressure on it."

"I'm going to give it another shove," Isabelle said. "At the very worst, it would simply break off, and then the astrolabe would be easier to manipulate."

"Maybe that's the whole point," Michael said. "If you're desperate enough to break the thing—your one connection to the father you never knew—then it's the right time to break it."

Holding the brooch in the palm of her hand, Isabelle put her thumb on the lizard and bore down as hard as she could, trying to shift it farther to the side. It didn't budge. She handed it to Michael.

"If we cooled it off, the metal would contract and the fittings would be looser," he said.

"Just give it a shove, old-timer," Hank said. Michael cocked his elbow and rammed his thumb onto the lizard. Suddenly it flew off into the air and the dreamcatcher fell to the floor of the pilot house. Michael grabbed it hurriedly and looked anxiously to see what damage he'd done. Evidently the surface of the brooch had been held in place by the lizard; once the lizard sprang loose the background material came off too, exposing several additional intricately incised dials. Engraved in a small bare panel in the middle were the following markings, hardly big enough to read:

43.8789° N, 103.4598° W

Michael rose to his feet and showed the others the interior of the dreamcatcher, trying to repress his excitement.

"You did it!" Isabelle exclaimed. She leaned over and threw her arms around him.

"I could just as easily have wrecked it."

"Desperate times call for desperate measures," Hank said.

"And now we know where we're going," Isabelle said. "Isn't that the point? I'm so relieved."

She paused, and then said a little sheepishly, "Michael, do you know where that is?"

"Not exactly," he laughed. "But I can tell you it's a long way from here. The hundredth meridian used to be referred to as 'where the West begins,' back when people inhabited the Dakotas and Nebraska, hunting buffalo or raising cattle and sheep. And the forty-ninth parallel marked the border between the United States and Canada in the days before the continent was fully federated. Wherever Dr. Grace located his lab, I'll bet it's on high ground south of the Canadian border and east of the Rockies. I'd be very surprised if we could get to such a place traveling on water."

"Well, let's hope we can. If we do, it won't be the first time we've been surprised recently by a turn of events," Isabelle said.

"That's true," Michael laughed. "And the good news is, we can continue east, navigating by landmarks while we wait for the stars to reappear—if they ever do. I've flown this part of the world many times while servicing the towers on Sandia Peak. I know quite a bit of the terrain, though of course it looks different from thirty-thousand feet. And I'm sure this old tub has an inertial navigation system on board that doesn't require external points of reference. Put those things together—"

"—and we'll probably run aground in a canyon some-where in Arizona," Hank quipped.

"Oh ye of little faith," Isabelle chided him.

"Don't you think we should tell Laura about all this?" Michael interjected. "How much does she already know? How much does she remember after all these years?"

"She's been through a lot, Mike. It sometimes seems that her sanity is bound up with her determination not to reveal what she knows—to anyone. At this point I can't tell if her silence is based on lingering suspicions or gnomic wisdom, but I firmly believe that her presence will prove to be invaluable once we reach our destination. She already loves me like a daughter, and she's warming to you as she sees you in action—sees the man beneath the technocrat. I suspect that if we told her what we've just discovered about the dreamcatcher, she'd just smile and nod."

14

Below deck the sounds of the storm had lost their edge and coalesced into a generalized ululating moan interspersed with metallic thuds and pings as the dilapidated vessel went head to head with the elements. Passengers huddled in corners and under the wreckage of slot machines and overturned blackjack tables. Isabelle had gathered together a cadre of women and men capable of treating the various injuries that passengers had sustained during the collapse of the city and the subsequent hasty departure of the casino barge. There were now no pressing medical emergencies to deal with. One-Eye had handed out blankets, and when Michael and Hank returned below deck they helped him distribute rations of food and water. The lights in the casino had been dimmed to conserve energy, and an almost pleasant calm had settled over the motley group of survivors. The jostling motion of the barge was now less violent, and to judge from the sound levels outside, the storm might well be dissipating slightly.

Would it be over soon? Not likely, Michael thought, and perhaps not even desirable. For one thing, they were going to need a good deal of water to reach their destination. And though the Blue Skies project had resulted in catastrophic loss of life, it was still impossible to predict what the balance of atmospheric gases would be once a new equilibrium had been established. Perhaps the chemical reactions that were playing themselves out would result in a more hospitable planet? One could always hope.

Michael was eager to return up top and take another look at the surroundings, and having distributed the packets of rations he made haste to do so.

Rain poured down as he stepped out onto the deck. To judge from appearances, they could be out in the middle of the ocean. But it was an ocean dotted with small flaming islands of methane hydrate that had been released from the ocean bottom by the earthquakes. The orange-blue flames were almost beautiful as they illuminated the lower reaches of the still lightning-filled sky.

It was clear that the worst of the firestorm was over. But the water remained turbulent, and examining some of the landmarks in the San Jacinto Mountains that he recognized from his experience flying above them, Michael was shocked to discover that the water level had risen several hundred feet! Considering the length of time they'd been afloat, that was far too much of a rise to be attributable to rainfall alone, no matter how virulent it had been. The thought sent a chill through him, as he realized that very few of the citizens of Los Angeles were likely to have survived such a deluge. Even those who had made it safely into the bunker might still be in serious danger of drowning or suffocating.

When Isabelle appeared at his side, Michael put those thoughts out of his mind and reached his arm around her waist. They stood in the rain, rocking gently together with the waves. Pointing off to the east with his free arm,

Michael said, "Can you see how far the water has risen? It's astounding."

Isabelle merely murmured a sigh. She was tired and strangely happy. She didn't want to think about water levels, and she hardly listened when Michael continued: "We've been witnessing an increase in earthquakes and tsunamis for decades now. It isn't merely the isostatic rebound caused by the glaciers that vanished eons ago, though that's still taking place. No, it's the rebound caused by the recent loss of the polar ice caps. A slight decrease in pressure as the crust returns to its former levels can lead to the opening of volcanic conduits, and once they're open, the magma corralled in the chambers beneath them will force them open farther."

"So, more underground volcanoes?" Isabelle replied.

"Yes, and also a lot more surface water. But no one could have calculated a change of this magnitude. And there's no way for us to tell how much more to expect. We'll be lucky if we don't get swamped by a tsunami before we're through."

"And when *will* we be through, Michael?" Isabelle asked dreamily.

"Our only hope now is to find Dr. Grace," Michael replied. "After that, we'll just have to see."

Down below, Amber was staring at the rat.

"What?" the rat said finally.

"Why are you still wearing a rebreather?"

The rat shrugged.

"It's okay to take it off now."

A tear appeared in the corner of the rat's eye. She wiped it away and then reached up to detach her rebreather. As she did so, Amber could see her leathery skin.

"My dad said all rats are bad," Amber said, "but you're not bad, are you?"

"Where there is hatred, let me sow love," the rat replied. "Where there is injury, pardon. Where there is doubt, faith. Where there is despair, hope. Where there is darkness, light. And where there is sadness, joy."

She looked down at Amber and said, "My mother taught me that."

"My mother's dead," Amber said.

"Mine, too."

* * *

The buttes and mountains crept slowly by, and so did the days. Rain continued intermittently, and vast swells often appeared from the west, lifting the vessel and propelling it forward at enormous speed. With the aid of dead reckoning supported by the ship's gyroscopes, Michael and Hank struggled to chart their position, day after day, but Michael was almost overjoyed when he recognized the gentle cone of Mount Taylor rising from the sea to the north—a valuable identification considering they lacked navigational maps. When Sandia Peak—the Watermelon—came into view, like an elongated dome that had never been inhabited, everyone breathed a sigh of relief.

"We're going to hang a left turn around that lovely lady," Michael said, struggling to keep the excitement out of his voice, "and then it will be smooth sailing for a while north along the Sangre de Cristos." He looked over at Hank, who was looking a little glum.

"What's wrong with you, buddy?" Michael said.

"You know what else I see," Hank replied. "I see the tower rising up into the clouds from the top of Sandia Peak."

"And it reminds you of our grand project?" Michael said. "How much do you really know about it?"

Just what I read in the papers."

"Well, let me tell you, the science was good. Damn good. But Mother Nature it turned out is not to be tampered with on such a grand scale. Call it hubris. At this point I guess I'm not feeling guilt so much as nostalgia. Is that an improvement?"

"Hubris, my eye," Hank replied. "Desperate times call for desperate measures."

"You said that already," Michael replied.

"Well, what's done is done," Hank said. "And I want to tell you fellas something. That tower we're approaching is one impressive piece of engineering. It goes so high I can't see the top."

"For one thing, the tower is six miles tall," Michael said. "No one could ever see the top of it. And it's impossible to see anything well in this thick atmosphere anyway."

"Really? Aren't you surprised by how far we're seeing now? The atmosphere has definitely been improving. I've even been seeing a few stars now and then at night."

"You know, Hank, I think you're right," Michael said. "I've noticed the same myself but chalked it up to wishful thinking. And if you're right, that's good news. It will be relatively easy for us to move north along the Front Range, but at a certain point we're going to have to head off northeast across the plains. When that day comes, having a few stars to navigate by will be a godsend."

"And when will that day come?" Hank asked.

"It's hard to say. We might be halfway to our destination, but we've been helped a lot by the swells coming from the west, and I suspect we'll lose most of that impetus as we begin to move north across Colorado."

"One thing's for sure," Hank said. "We going to have to tighten up on rationing."

* * *

As rationing grew more severe, some passengers began to question the wisdom of having abandoned the city, while others relished the empty days and the relative calm onboard. For some, it was like being in prison again, though the guards were nicer and the food was worse. No one really wanted to go back. Everyone had claimed a corner or a table under which to set up a personal space. Some were quite creative in their decorative schemes, using outlandish banners and raffle prizes that had been purchased in bulk by the casino years earlier. No one spent much time out of doors, though Isabelle set up a yoga class on deck to keep people moving and healthy.

One evening, as the barge droned on toward the east, with mountain ridges like the fins on a dragon's back spreading across the northern horizon, Michael and Isabelle were examining the evening sky when they were joined on deck by Laura. She had often seemed exhausted and weary, but on that night a twinkle had returned to her eye. She stood beside the couple for a few minutes in silence, and then she said simply, "I want to tell you something."

"What is it?" Michael responded, trying not to sound overly excited.

"I want to tell you how it happened."

"What happened?" Isabelle asked.

"We were in the laboratory at the zoo, your father and I." She spoke slowly, choosing her words carefully. "Conditions had been deteriorating for some time, but our security system was still functioning. We were extracting DNA, I don't remember the species. You'd think I would," she said in a vaguely bemused tone.

Fearing that her mother would lose the thread, Isabelle said hurriedly, "It's not important. That was a long time ago. But what happened then?"

"It *was* a long time ago," Laura said. "I was carrying *you* at the time. And just look at you now!" She paused, then continued. "They breached the entrance. We heard them coming. Saw the monitor. Didn't have much time."

"Who was coming?" Michael said.

"President Reyes and his men. Of course, he wasn't president then, but he'd already developed a reputation for efficiency and daring. He knew how to get a job done."

"So that's how they captured you?"

"They tripped the sirens on the way in and the doors closed, but they'd come prepared and they blew them apart. It was a well-coordinated raid. We didn't have much time to gather together the specimens. Many were already safely stored away, of course."

"Stored? Stored where?" Michael asked.

"Many had already been deposited at our lab. And other specimens were in the ARK. That's the titanium chest we used to transport things. That day James lowered it into one of the escape pods and asked me to get in. There were two pods. I said, 'You're coming, too,' but he refused. Claimed there wasn't room. Wanted to save the baby. So I tricked him. I had to do it."

Laura paused again, and it seemed that she was still feeling slight pangs of remorse because of what she'd done on that fateful day.

"How did you trick him?" Isabelle said.

"I told James that I'd forgotten how the thing worked. He got in to refresh my memory, show me what to do. Once he'd identified the controls that would activate the pod and get it going on autopilot toward our little airport, I reached over his shoulder and activated the pod. I almost lost a limb as the canopy to the cockpit slammed shut. He'd already handed me the dreamcatcher, said it would guide me to a place where we could cache the specimens indefinitely ..."

"So he got away safely with the ARK, and you were left to face the authorities."

"That was the better solution, I thought. There was pounding and shouting. Reyes was smart enough not to blow up the entire lab, so it took them a while to get in, and they didn't find anything of value when they finally arrived."

"They found you," Michael said.

"A lot of good *that* did them," Laura chuckled.

"Do you know where the lab is now?" Isabelle asked gently.

"I've been there. I'd recognize it if I saw it. Up in the hills."

"What hills?" Michael said.

"There are so many hills," she said. Then she sunk back into herself, and they both knew she'd withdrawn from the conversation.

* * *

With the passing days Michael grew more gaunt, like everyone else; the beard enhanced his boyish looks. His relationship with Isabelle had grown deeper, though they made every effort to be discreet. Their quest had taken on the dimensions of a dream, a hallucinatory expedition into the unknown. There was nothing else to grab hold of, and their faith in the mission had grown stronger as its impracticality grew more obvious.

Early one morning Michael was awakened by the sound of something on deck. Delicately disentangling himself from Isabelle's pleasant embrace, he slid out from under their table-home and headed up the steps. As he emerged on deck, dawn appeared as a smudge of gray in a storm-black sky. Laura was kneeling in prayer at the gunwale of the gently bobbing ship. Michael knelt beside her and closed his eyes.

Without opening her eyes, Laura said, "Hank is right. The air is getting cleaner."

"That's what we'd hoped would happen, though the cost has been infinitely greater than we anticipated," Michael replied. "The rain's been dissolving the sulfides and hydrocarbons out of the atmosphere. The intervals of sunlight have given the batteries a boost, and we've even been seeing a few stars at night."

"Mother Earth's finding a way to heal. There's an indwelling motion toward balance, and life."

"And the casino barge has also found a way to stay afloat."

"Thanks to you and Lance."

"But it won't mean much if we don't find the ARK. Which may well be lying under a thousand feet of water...if it still exists. Do you think we'll find it?"

"I believe that we each have a part to play in a master plan," Laura replied. "It's written in every book, in every religion—the importance of doing that one thing you were put on this earth to do. The hard part is figuring out what that something is. Harder still, perhaps, to stick with it once you find it."

She opened her eyes and looked at him. "But deep down, we already know. The day you promised your son you'd find a way. On that day, you knew."

"Seeing the disasters of the last few days and weeks has made me far more aware of what's at stake. And far more doubtful whether we'll succeed. Do you mind if I pray with you?"

"I pray when I want to speak to God, and I meditate when it's time to listen."

"Which were you doing just now?"

"Listening. But you should do what you need to do."

Michael nodded, then closed his eyes. Laura returned to her meditations, and her countenance became enveloped in a soft blanket of peace.

*   *   *

The morning finally arrived when the mountains seemed to be dropping into the sea and also veering off to the west. It was time to leave the shoreline behind, venture farther out into open water. Michael had never flown in the vicinity, but based on the coordinates in the dreamcatcher, there could be little doubt they were headed for somewhere in the Black Hills. In recent days Hank had been able to spot a few constellations just before sunrise, and using the astrolabe to determine how many degrees the sun was above the horizon, he was able to approximate the latitude. But Michael was more worried about the longitude. They were still drifting east with the currents, but if they strayed too far in that direction they'd end up traversing the vast plains of Saskatchewan! The ship's compass became an invaluable tool, and an occasional sighting of the North Star was always reassuring.

Four days later, Michael made a discovery. He was in the pilot house, once again just at dawn, when a shocking image met his eye: Teddy Roosevelt's glasses!

"Mount Rushmore! This is it!" he said to himself excitedly. Then, calming down a little, he said, "Anyway, we're close." He called down to Lance and instructed him to rouse whomever he could. Soon a small party was standing on deck, admiring the foreheads of the Founding Fathers.

Seeing Isabelle, Lance, and the others standing togeth-

er in a rare patch of genuine sunshine, Michael was struck by how weak and gaunt everyone looked. Still, there was excited talk about going ashore. People shared their opinions about the risks involved, and how difficult it would be to locate a proper landing site.

Suddenly Isabelle exclaimed, "Look over there!"

"What is it, dear?" Laura said, fearing that her daughter had seen something threatening.

"I see something metallic—a brilliant flash of light," Isabelle said, pointing. "Can you see it? Is that the fuselage of an airplane?"

Michael spotted the object immediately, sitting on the side of the cliff, and a shiver ran through his body. It looked old, but how old was it? No one had been out this way in decades, unless the rats were far more widespread than he imagined. Had a plane crashed? Whose was it?

"That's totally cool," Lance said. "Let's check it out."

A sadness came over Laura as she stared at the distant object. In her heart of hearts, she knew it was her husband's plane. Soon the hope she'd been living with for decades would be realized—or dashed.

Seeing her downcast expression, Lance went over to her, pointed to the enormous faces half-visible above the water, and said lightheartedly, "I recognize George and Martha, but who are the other two?"

She gave him a woeful look but didn't smile.

Michael caught her eye. "Is this it? Are we here?" his upturned brow queried. She nodded almost imperceptibly.

An hour later Hank and One-Eye were tying off the rope on the trunk of a long-dead juniper tree near the base of Washington's head.

"That'll hold her," Hank said. "Let's rig up a gangplank." Soon Michael, Isabelle, Laura, and Lance were disembarking with supply sacks hoisted over their shoulders.

"Hey, Hero," Hank called out. "I found this down below." He handed Michael a flare gun. "If you run into trouble, just take it. It's a jungle out there," Hank said, sweeping his eyes across the rocky and largely barren hillside. "Besides, you might twist your ankle."

"Thanks. We'll make every effort to be back before nightfall. I doubt we'll need anything like this."

"Just take it. It's a jungle out there," Hank said, sweeping his eyes across the rocky and largely barren hillside. "Besides, you might twist your ankle."

On that note, Michael, Lance, Isabelle, and Laura headed up into the hills, following what might once have been a well-worn path. Months of rain, and perhaps decades of disuse, had largely obliterated it.

When they reached the ridge the path deteriorated and Michael paused, then stepped aside to let Laura take the lead. An aura of mystery, almost of reverence, enveloped the party as they trudged across the still-wet and often slippery terrain.

"I think I know where we are," she said at one point, pausing to catch her breath. She didn't sound confident, but she wasn't asking for advice, either. She was groping her way, guided by a collection of memories and also, it seemed, by an array of forces that continued to draw her across the landscape. Lance was thinking they could just as well have chosen a less demanding route to the same spot, but for once he knew enough to keep his mouth shut.

Continuing up the canyon, the party came to a narrow defile. The path continued to climb, and when they reached the top, a new vista opened up below them. "This is it. We're very close," Laura said in hushed tones.

Michael had already begun to notice odd patches of lichen on the rocks, a salt bush that seemed to be doing unusually well, and several paths in the sand that might have been cut by lizards or snakes as they made their nocturnal rounds across the desert floor.

At one point early in their descent into the next valley Isabelle paused to examine a robust field of plants bearing tall stalks on the hillside. "Look at this!" she exclaimed.

The others paused and looked at the nondescript scrubs.

"That's a very nice weed patch, Isabelle," Lance said. "What about it?"

"These are sunchokes. Jerusalem artichokes!"

"We're a long way from Jerusalem, dear," Michael said.

"That's just a fractured version of the Italian 'girasola,' which means 'turning to the sun.' The flowers of this plant do that. Of course, so do lots of other plants. But the important thing is, the roots of these plants are very rich in nutrients. They're also known as ground potatoes. If I didn't know better, I'd think they'd been planted here."

"You certainly know your plants, Isabelle," Lance said. "Where does all that come from, anyway?"

"It runs in the family," she replied, looking fondly at her mother. As she did so, Michael noticed that the scar on her face had disappeared. He touched the area where it had been and said quietly, "it's gone."

Isabelle raised her hand to her face and slowly ran her fingers across the area where the scar had once been, mystified by the unexplained healing.

"Maybe these plants were planted here," Michael said, with a tinge of excitement in his voice. "If Dr. Grace is anywhere in the vicinity, he'd have long since come up with an ingenious scheme for staying alive."

"I guess we'll soon find out," Laura said. "Shall we proceed?"

As they continued down into the valley, it soon became obvious to everyone that something unusual had taken place here. The vegetation was as verdant as an arid locale could be, and on several occasions Michael thought he heard the chirp of a bird. It was like an oasis, and simply walking through it put everyone in a good mood.

Soon the party came upon the entrance to a cave. Michael noticed that Laura misted up briefly when she saw it, and even smiled a little. Michael was going to make a little speech, preparing everyone for the worst and cautioning them that some dangerous creature might be inhabiting the cave. But he kept quiet, and the party followed Laura in reverent silence as she fearlessly entered the shadowy opening in the rock.

"This has to be it," she muttered to herself, and her step quickened. Lance had produced a few flashlights from his pack and handed them around, and soon beams of light were dancing off the rock walls in every direction. As Laura methodically proceeded up the dark orifice, Lance noticed something strange: her figure was silhouetted by a glowing light ahead of her, as if sunlight from the outside world had found a way in through a crevice in the rock. Soon the flashlights were no longer necessary.

At the same time Michael began to notice the sound of a steady drip, as if groundwater was percolating through an aquifer nearby. The air was also becoming fresher—almost sweet. And Michael began to notice other strange things: lichens on the walls of the cave, and moss on the rocks at his feet.

Following the path through the cave, they eventually came to a passage where the ceiling rose and the floor opened out. Michael could see rays of sunlight stream down across the cavern ahead of them from somewhere far above. Here the water had formed into pools surrounded by mosses, grasses, succulents, and even a few hardy wildflowers.

"This can't be real," Isabelle said with a gasp of bewilderment tinged with excitement.

"Are these organisms from the ARK?" Michael said in an awestruck tone, hardly believing his eyes.

Laura began to answer but then gave a cry and put a hand to her mouth. Out of the shadows behind them a figure was emerging.

A faint, croaking voice said, "Welcome, dearest."

With a little shriek, Laura rushed into the man's arms. They embraced for a good long while, rocking gently back and forth with their eyes closed, murmuring things to one another. The rest of the party stood in respectful silence as they embraced, ogling the lush terrain all around them.

Finally Laura turned to her daughter and said, "Isabelle, this is your father."

Isabelle felt a tingle run up her spine. Not sure how to approach the frail figure standing in front of her, she held out her hand, but Dr. Grace pulled her, too, into a loving embrace.

"Oh, my little girl," he murmured.

During this moving but highly personal reunion, Michael had been furtively examining the cave, and he'd noticed a metal box in the shadows up against the rear wall. When Isabelle and her father turned once again to the party, everyone stood silently by the pool, unsure of how or when to intrude upon this emotional scene.

Dr. Grace cleared his throat—he hadn't been doing much talking in recent years—and then said, "The weather's been strange lately. Lots of wind and rain. Lightning. Strange gases. The atmosphere seems hyper-charged in a way that I can't explain. Though it's getting better now."

Michael said, "I can fill you in on the details surrounding that apocalyptic debacle. I was the scientist in charge of making it happen."

"Now just a minute," Isabelle interjected. "Michael is referring to a worldwide initiative called Project Blue Skies designed to reduce CO_2 in the atmosphere. Every detail of such a large-scale project can't be tested ahead of time, and the atmospheric reactions involved didn't play out quite like he'd planned. But the storms and flooding you witnessed here, while very destructive in some locales, also cleared pollutants from the atmosphere, reduced CO_2 levels, and gave us hope again."

"I would appreciate hearing more about these things a little later," Dr. Grace said, nodding deferentially at both Michael and Isabelle and cracking a slight smile. Perhaps his daughter was in love? If so, then it seemed she had made a good choice.

He paused briefly to clear a lump in his throat, sighed, and took a deep breath. Then he said, "It's hard to believe you're all here. I am overjoyed to be once again in the company of loved ones."

He paused, tried hard once again to get a grip on himself, and said, "I'm sure Laura brought you all here for a reason. And I expect that you already know a little something about what I've been doing here—preserving biodiversity. Perhaps, with help from all of you, and under a bluer sky, we can continue our grand mission on a larger scale."

Michael said, "I don't know much about your project, Dr. Grace, but Laura has explained a few things to me, and I've read about your early career at the zoo just like everyone else. I'd like to know more."

"That's an understatement," Lance said with a chuckle.

"We have a few supplies and technical resources back

on the boat," Michael continued. "But also quite a few malnourished passengers."

"I think there are resources hereabouts to deal with that problem," Dr. Grace said, "though we wouldn't want too many people tramping way up here just yet. Eventually, the added labor will be essential to improving our stores of edibles."

"Let's discuss briefly what we have back on the boat that might be useful," Michael said. "Then I'd like you, Lance, to go back and reassure everyone that we haven't fallen off a cliff. You can pick one or two individuals to help bring the things we need back up here, and others to gently maintain order and reassure everyone we'll soon be able to tell them what's going on and develop a realistic timetable for our activities. Does that make sense, Dr. Grace?"

"Good plan. Now I'd like to know what kinds of equipment you have on board."

"Before I head back, maybe you ought to tell me a little about what's going on here. Just in case someone asks," Lance said.

"In brief, I began a program many years ago to collect and preserve organisms from every phylum and as many genera and families as I could—their DNA. You might even have read about such things in a newspaper, though that was decades ago. Things were already beginning to look bad in those days. Politically, I mean. Need I say more? I still have that material, though it's been difficult keeping the necessary equipment running over all these years. With your help, as conditions improve, we may be able to begin repopulating the earth. As I'm sure you know, once a community of organisms gets established, one thing leads to another. Who would have believed when Krakatoa blew its top in 1883 that within a few short years it would be teaming with hundreds of life-forms again?"

* * *

In the days and weeks that followed, bolstered by new equipment and knowledgeable personnel, Dr. Grace continued the long and tedious task of cell subdivision and replication, with the end in mind of bringing back from extinction an array of species that had not thrived on earth for decades. They followed a careful plan, focusing on plant, insect, and rodent species that, once revived, would support one another naturally within a given ecological niche.

Michael proved to be adept at the work of regenerating the DNA and stimulating cell subdivision and replication. He found the hands-on element of the work a refreshing contrast to the highly technical and highly speculative aspects of the Blue Skies project. He and Dr. Grace worked well together.

One afternoon, everyone gathered in the makeshift lab, eager to witness an especially satisfying milestone in their work. They were all aware that an ecosystem requires organisms of every size and description to prosper, including the gnats and fungi, but today they were hoping to revive something special. Michael was wearing a pair of halogen ophthalmic glasses as he worked with Dr. Grace standing by his side. Michael concentrated on the monitor as the instrument he was holding moved slowly closer to the cell wall. As the probe penetrated the cell, it began to divide: two cells, then four, then eight. Life had begun. Dr. Grace exhaled. Michael flipped up his goggles and grinned. Dr. Grace grinned back as he spread his arms wide and said, "Danaus plexippus."

Isabelle gave Michael a big hug as he said, "And thus, the monarch butterfly emerges from extinction." There was cheering and high-fives all around.

* * *

Meanwhile, back in Los Angeles, life inside the bunker remained secure, if claustrophobic. A system for the orderly distribution of rations was put in place, and soon the screams and cries of the recent rebellion were replaced by the anodyne hum of a generator and the shrieks of children playing together in the bunker. Alex was put in charge of technical maintenance, and he put together a crew that included Deborah and also Maxwell, who had proven to be as good with machinery as he'd claimed. The ventilation system posed a continual problem, with fans succumbing to the damp air outside and duct filters clogging. It was also important to monitor conditions outside the bunker. Although the city dome was now essentially useless, plans were already underfoot to rebuild on a smaller scale, if and when the flooding subsided. People couldn't stay cooped up in the bunker forever.

During those first weeks Alex found his thoughts returning again and again to President Reyes and Michael Costello. He had scorned the president for hoarding supplies to support an elite population at a time when rationing was growing ever more harsh in the city for the common citizen. Yet here Alex was, taking advantage of those spoils and grateful that they would not be running out anytime soon.

And what about Michael? Why hadn't he made it to

the bunker? For that matter, why had the Blue Skies initiative gone so terribly wrong? Was that final, perhaps over-cautious blast of energy the factor that had tipped the chemical reactions in the wrong direction?

These were things he would probably never know.

But one afternoon, while he was tinkering with an uncooperative thermostat, Stephanie came running up to him.

"Alex!" she shouted. "You've got to come."

"Come where? What's going on?"

"A message on the mainframe," she shouted over her shoulder as she hurried back the way she'd come. Alex followed her down the overcrowded halls of the bunker to the computer room.

"This came in a few minutes ago via satellite," she said, still a little out of breath.

"Why all the mystery and suspense?" Alex said, a little annoyed in spite of himself.

Lieutenant LeBlanc, who was standing at the keyboard, retrieved the message.

Mike's face appeared on screen. He seemed to be in a very odd place—certainly not a well-lit room or laboratory. He looked tired and gaunt, but his face was beaming. All he said was, "Wanna help us save the world?"

Tears filled Alex's eyes as he sat down in front of the screen. With shaky hands, he typed, in capital letters: YES.

Acknowledgments

Many people contributed to the development of this book. The authors would like to thank our editor, John Toren, our illustrators, Bly & Rowan Pope, and our front cover designer, Joan Holman. Among the many scientific experts who helped us shape the story line, we owe a special debt of gratitude to Tom Overton (retired NASA Shuttle Processing Manager), Dean Pannel (concept developer & futurist), Eric Milbrandt (Marine Laboratory Director of the Sanibel Captiva Conservation Foundation), Craig Tepper (research associate and Chair of the Cornell College Department of Biology), Miriam Scanlon, Barry ZeVan, and Michael Bondi.

Lee E. Frelich is Director of the University of Minnesota Center for Forest Ecology. His research examines the impacts of climate change, habitat loss and invasive species on forests.

John B. MacDonald is a conservationist, writer, and entrepreneur. His main interest is the long-term collection and preservation of DNA from plant and animal species facing imminent extinction.

Made in the USA
Middletown, DE
22 July 2017